B-17 FLYING FORTRESS

Text by ROGER FREEMAN / Illustrations by RIKYU WATANABE

Crown Publishers, Inc.
New York

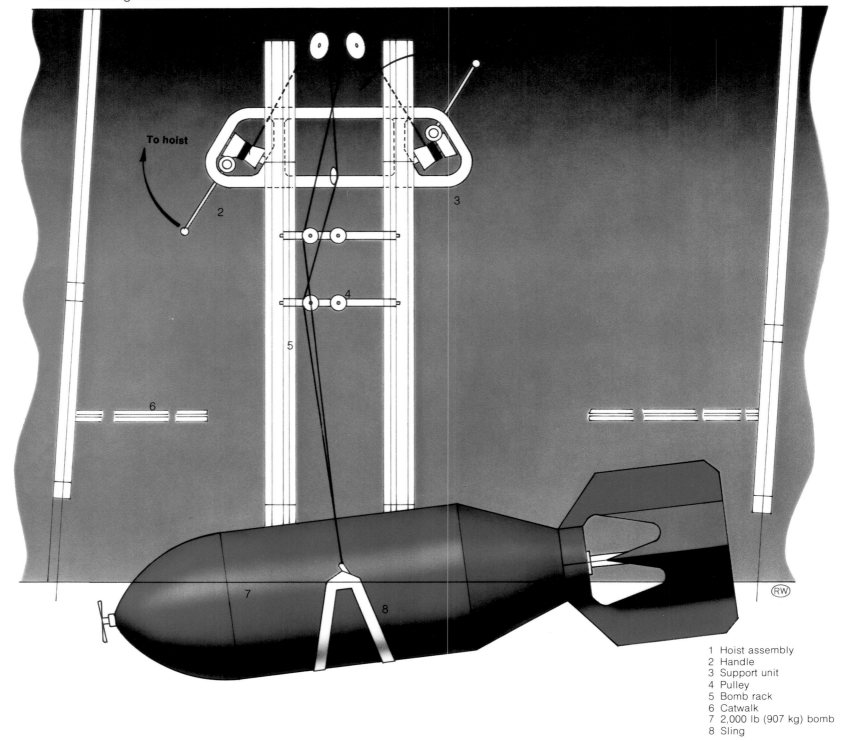

To hoist

1 Hoist assembly
2 Handle
3 Support unit
4 Pulley
5 Bomb rack
6 Catwalk
7 2,000 lb (907 kg) bomb
8 Sling

This book was designed and produced by Wing & Anchor Press,
a division of Zokeisha Publications, Ltd.
5-1-6, Roppongi, Minato-ku, Tokyo 106/123 East 54th Street, New York 10022.

© 1983 by Zokeisha Publications, Ltd.

First published in the U.S.A. by Crown Publishers, Inc.,
One Park Avenue, New York, N.Y. 10016.

Printed and bound in Japan. First printing, January, 1983.
Library of Congress Cataloging in Publication Data

Freeman, Roger Anthony.
  B-17, Flying Fortress.

  1. B-17 (Bombers) I. Title.
UG1242.BF73 1983   358.4'2   82-23631
ISBN 0-517-54985-9

Newly rolled-out Boeing B-17F-40-BO Flying Fortress in flight.   (BOEING)

Legend

Such is the fame of the B-17 Flying Fortress that stories of its exploits have already become legendary. Blurred by time and hearsay, it is not always easy to separate fact from fiction and assess just what contribution the aircraft made to the Second World War. Few would dispute that bombing played a major part in inducing the surrender of the Axis powers; and in this respect it can be stated that two-fifths of the total tonnage of bombs dropped by the US Army Air Forces were delivered to their target areas by Fortresses. To achieve this, crews had to fight their way through the enemy's airspace, enduring the hazards and discomforts of high altitude flying to minimize the effect of anti-aircraft fire, yet be alert to man the guns to ward off enemy fighters. The aircraft had to be large to permit heavy bombs to be lifted and carry the fuel necessary to fly long distances. In view of the enemy defences on the ground and in the air, they were of necessity heavily armed and armoured – in fact, they were literally flying fortresses. The name itself has a magic ring and supports the legend – but here is their story in fact.

The bomber had its origins in the US Army Air Corps' doctrine of high altitude precision daylight bombing to attack strategic targets and deny an enemy the means to sustain their opposition. By establishing air power as a major force in warfare the air leaders were hopeful that they would gain autonomy for their own service, as Britain had done in 1918 with the establishment of the Royal Air Force. This was an expectation not to be fully realized until 1947 – when the Fortress was already outdated. When the potential of a new Boeing bomber was first recognized in the mid-1930s, it was seen as the means to foster this doctrine, and so the Fortress spearheaded the American offensive in Europe and remained operational throughout the Second World War.

Lineage

The genesis of the aircraft can be traced to a design project originated by Boeing in the summer of 1934. At that time the future of the company seemed uncertain as the US government had forced the break-up of United Aircraft, an aeronautical consortium which included engine, airframe and component manufacturers plus an airline. The government viewed this as a cartel and, indeed, Boeing had an order for 60 of its Model 247 passenger transports from the consortium airline, United. This was seen as unfair competition and legislation ensured that Boeing would no longer have the benefit of a guaranteed market for its products. The company was experienced and enterprising, as were most of its competitors, but the national economy was severely depressed and few commercial or military orders were being placed. Despite this the period witnessed a rapid advance in aeronautical technology with Boeing among the leaders.

William Boeing had started building seaplanes at Seattle, Washington state, in 1916 and his business prospered with orders from the US Army and Navy and for civil transports. In 1930 the company flew an all-metal low-wing monoplane, the Model 200 *Monomail*, featuring a simple retracting undercarriage – an innovation for the period. The substantial low-wing configuration, employing tubular-truss spar construction, featured in following designs; these were the progressive but unsuccessful Model 215 twin-engined bomber (later designated YB-9) and the successful ten-seat Model 247 twin-engine airliner. All these aircraft, powered by Pratt & Whitney air-cooled radial engines producing top speeds ranging from 150–200 mph (241–322 km/hr), outpaced the biplanes that predominated at that time. The expertise gained in producing these aluminium structures encouraged Boeing to pursue designs for even larger monoplanes when a requirement arose.

The US Army's Materiel Division pursued a policy of circulating requirement specifications for new aircraft to the industry, inviting manufacturers to submit entries for competitive evaluation. Normally the cost of producing a prototype was met by the manufacturer. In the biplane era of wood, wire and fabric this system did not usually impose a great strain on company resources. However, with the advanced technology involving the widespread use of metal in airframes not all companies could sustain the investment necessary for such a gamble. For the US Army Air Corps there were tight military budgets and only a small number of active flying squadrons permitted, yet despite this they continued to make efforts to promote new technology. While many senior officers nurtured an interest in a bomber capable of long-range strategic operations, this was no part of official Army policy and certainly not in line with the then isolationist stance of Congress. Development of a long-range bomber could only be pursued if seen as a necessary means of defending the United States' overseas possessions or protectorates, such as the Hawaiian and Philippine islands.

In early 1934 the Army offered a contract for the design, wind tunnel testing and technical data of an experimental long-range bomber, appropriately designated XBLR-1 (Experimental Bomber Long-Range Number 1). The Boeing company, seizing the chance to further their monoplane designs, tendered. Their submission was accepted in June 1934 which led to an order for a prototype a year later. This was truly a giant aircraft, spanning nearly 150 ft (45.72 m) and weighing 20 tons. Difficulties encountered during its construction delayed the first flight of the aircraft until October 1937, by which time it had been redesignated XB-15.

While design of the XB-15 and tail-off work on fighter and transport production kept Boeing in business, their future did not look bright unless a substantial airline or military order could be obtained. The company therefore decided to work up preliminary designs for two multi-engined aircraft types – one military and one commercial. It was expected that later in the year the Army Air Corps would issue requirements and invite bids for a bomber to replace the Martin B-10 currently serving in bomber squadrons. This would be of the then conventional size, which later became the "medium" classification. The manufacturer of the successful prototype could expect an order in excess of 100 aircraft, a prize well worth the gamble. Paralleling this design would be a commercial transport utilizing as many components of the bomber design as practical. The Boeing design team under Claire Egtvedt started work in the summer of 1934 and included several features proven in previous aircraft in their preliminary drawings. Understandably, work on Model 294, the large experimental bomber that was to become the XB-15, was a major influence and the new bomber project, Model 299, included several similar features and looked rather like its scaled-down version.

The anticipated "Circular Proposal" for a new bomber for the Air Corps was received early in August that year. Stated requirements included a top speed of 250 mph at 10,000 ft (400 km/hr at 3,050 m) carrying a "useful load", and a 220 mph (354 km/hr) cruising speed at the same height at which endurance was to be 10hr. The aircraft was to reach this operating altitude in 5min and it was also required to maintain level flight carrying its useful load at a minimum of 7,000 ft (2,130 m) with any one engine out. In view of this final requirement and the overall high performance, Boeing engineers considered using four instead of two engines to power their entry. The specification had stipulated only "a multi-engined 4 to 6 place land type airplane". While a twin-engine design was the general expectation of the Air Corps officers drafting the specification, Boeing's enquiry on this point brought assurance that there would be no objection to the use of four powerplants.

Prototype

Boeing was authorized to design and build the prototype on 26 September, 1934 with an initial funding of $275,000. Subsequently additional funds had to be found and well over double the original sum was expended before the prototype flew. Egtvedt's design team was steered by the Project Engineer, E. Gifford Emery, and his assistant, Edward C. Wells, who had a major influence on the design of the aircraft. In basic configuration Model 299 had a wingspan of 103 ft 9 in (31.6 m) and would be powered by four Pratt & Whitney Hornet radials developing 750 hp each. The wing utilized the extremely robust tubular form strutting developed by Boeing and additional strength was built in by the use of a corrugated underskin. Within the wing, between the two engine nacelles, aluminium fuel tanks were sited. A 68 ft 9 in (20.95 m) long streamlined fuselage was divided into five compartments, the centre-section housing up to 4,800 lb (2,177 kg) of bombs in two vertical racks. Both mainwheels and tailwheel were retractable although part of the tyres remained exposed. An innovation was the use of the new Hamilton Standard constant-speed propellers which offered automatic pitch control to suit various power and altitude requirements. In contrast to the open defensive gun positions found on many military aircraft of the day, four closed gun cupolas were provided, their teardrop shape contributing to the smooth lines of the aircraft. The nose gun installation was also fully enclosed and the whole Plexiglas nosepiece could be turned to extend the field of fire of this weapon. A bombardier's sighting panel was installed in a rather ungainly fold in the underside of the nose. Side-by-side pilot and co-pilot dual controls in the cockpit were common to all Air Corps twin-engine bombers, so to facilitate four-engine operation a novel arrangement of throttle levers allowed either pilot easy manipulation of one, two or all engines with one hand. Features were added and changed during design and construction, despite the urgency of the aircraft being ready on time to compete in the evaluation trials at the Air Corps' experimental establishment at Wright Field, Dayton, Ohio.

Origin of the Famous Name

Model 299 was first "rolled out" for public display on 17 July, 1935. Boeing promoted it as the world's first all-metal four-engined monoplane bomber, but press representatives viewing it that day seemed more impressed by its sheer size and streamline. One reporter, Richard Williams of the *Seattle Daily Times*, fascinated by the prominent gun positions, was moved to write of a "flying fortress" in his report. An accompanying photograph captioned "15 ton flying fortress" was brought to the attention of Boeing public relations men and at a later date the term Flying Fortress was adopted as a company registered name. The appeal of the name lay in the United States' defensive posture as a fortress, alert to meet any attacker.

Eleven days after its first public appearance, Model 299 made a first test flight with Leslie Tower in the pilot's seat. The aircraft behaved well in this and subsequent flights, the only major problem being tailwheel oscillation during taxiing. On 20 August the bomber was flown from Seattle to Dayton for evaluation by the Air Corps. The flight, made in 9hr 3min averaging 233 mph (375 km/hr), no doubt impressed the examining officers at Wright Field. Even so, top speed was not as high as Boeing expected although it was considerably better than that of two other prototypes – from Martin and Douglas – also being evaluated. The Boeing was superior in

almost every aspect except price, which was more than double that of its competitors. For a small production run the Martin 146 entry was priced at $85,910; the Douglas DB-1 was offered at $99,150; but the Boeing 299 was $196,730.

On 30 October, towards the end of the evaluation programme, disaster struck. The Boeing, after taking off from Wright Field, immediately went into a steep climb, stalled and just failed to level out before crashing into the ground. The pilot, Major Ployer Hill, chief of the Wright Field Flight Test Section, was killed and the Boeing test-pilot, Leslie Tower, died from injuries received. Although the forward section of the aircraft was almost completely destroyed by fire, the tail suffered little damage and revealed the cause of the accident. One of the innovations of Model 299 was a system of control surface locks which could be operated from the cockpit. These locks prevented damage to ailerons and elevators from wind gusting when the aircraft was parked. Evidently Major Hill had failed to release the lock control and neither he nor Tower noticed this. The tragedy was a major setback for Boeing and there was little consolation to be had from the fact that the crash was due to human error and not to some fundamental weakness in the design.

The Boeing Company had committed the major proportion of its resources to the building of this aircraft and, as the outstanding tests could not now be completed, the contract would be lost to one of the other competitors. The superiority of Model 299 was unchallenged, but the fact that the Army could have two Douglas DB-1s for every Boeing purchased was an important consideration. The crash of the Boeing entry resulted in the decision late in the year to give Douglas a contract for 133 aircraft, to be known as the B-18, based on the successful DC-3 transport. All was not lost for Boeing as they were the recipients of an order for 13 Model 299s, to be designated YB-17 for service evaluation. The contract, issued on 17 January, 1936 for $3,823,807, covered the 13 aircraft, an additional airframe for static tests, spares and back-up. This was sufficient to keep Boeing solvent and their factory occupied for two years. Additionally, contracts were

Model 299 parked in front of the final assembly hangar at Boeing field in Seattle on 16 July, 1935. (BOEING)

YB-17 of 96th Bomb Squadron.

(USAF)

received from airlines for individual examples of the transport design utilizing the wing form and empennage of the Model 299, plus a large flying boat transport embodying several components similar to those of the earlier XB-15. While at this date Boeing had no substantial production orders, the future began to brighten.

Prior to the first flight of a service test Flying Fortress on 2 December, 1936 the official Army Air Corps designation was changed from YB-17 to Y1B-17 to indicate that the aircraft was specially funded. In practice the revised designation was rarely used and the aircraft were generally referred to as YB-17s, even on official documentation. The Y1B-17s, although externally very similar to the Flying Fortress prototype, incorporated a number of changes. The most important was the substitution of Wright R-1820 Cyclone engines for the Pratt & Whitney Hornets. Although a similar nine-cylinder single-row radial engine, also originating in 1927, the development potential of the Cyclone was greater; the R-1820-39 version, selected for the Y1B-17, was rated at 930 hp for take-off as against the Hornet's 750 hp.

Another notable difference between the Y1B-17 and the prototype was to be found in the main undercarriage. Model 299 had hoop-type legs which made tyre changing extremely difficult as the whole axle had to be dropped to remove the wheel. To simplify matters the landing gear was redesigned to feature a single oleo-type leg. The mainwheels, however, still retracted forward and up into the nacelles, remaining partly exposed. Rubber de-icer boots were fitted to the leading edges of wings and tailplanes. These were pneumatically operated with their inflation dislodging ice accumulation. The aluminium-covered landing flaps were fabric-covered on the Y1B-17 and there were several minor internal changes to vacuum pump equipment, fuel and oil tanks and instrumentation.

Early Service

The first Y1B-17 was delivered to the Air Corps in January 1937 and the 13th early in the following August. That same month 12 of the bombers were sent to an elite bombing unit of the Air Corps, the 2nd Bombardment Group based at Langley Field, Virginia. This Group was under the control of General Headquarters Air Force, created in 1935 as a special air strike force, having equal status to the Air Corps, while being also a corporate part of it. This anomaly nevertheless provided General Headquarters Air Force with the opportunity to develop a bomber organization with strategic capability, in spite of the presence of many senior officers in the Army hierarchy who opposed any tendencies towards developing separate missions for the Air Corps. They

firmly believed that all Army aircraft should be committed to supporting ground forces and be subservient to them. General Frank Andrews, commanding General Headquarters Air Force, together with many of those within his command, believed in the promotion of an independent role for air power. They were also wise enough to appreciate the opposition which lay not only in the Army, but also with the US Navy and certain factions of government.

The doctrine of strategic bombardment had been quietly nurtured for some time, although the idea of long-range bombers destroying an enemy's war industry was not new – Britain's Royal Air Force, having been instituted in the latter stages of World War I, was such an example. With the arrival of the Y1B-17s many officers within General Headquarters Air Force saw that at last there was an opportunity to investigate and develop strategic bombing. In view of the critics who deplored the expenditure on these large advanced aircraft, every effort was made to prevent accidents that might draw adverse publicity. On the other hand every opportunity was taken to show off their acquisition to good effect. In what was a carefully orchestrated publicity campaign, flight records achieved by the Y1B-17s were brought to public notice and in 1938 goodwill missions were flown to Argentina and Brazil. The publicity surrounding an incident where three B-17s intercepted the Italian liner *Rex* over 700 miles (1,130 km) out to sea as a navigational exercise, brought unexpected results when the US Navy demanded that the Air Corps restrict its activities to within 100 miles (161 km) of the shoreline.

In developing their bomber ideas, Air Corps officers promoted operation from high altitudes where, it was reasoned, a bomber aircraft would be at less risk from intercepting fighters and anti-aircraft artillery. To operate at altitudes where air density was lower, engine supercharging was necessary. The additional Y1B-17 airframe built for static tests had eventually been upgraded for development flying. It was this machine, known as the Y1B-17A, that was fitted with exhaust driven turbo-superchargers. Such boosting had been featured on earlier experimental and service aircraft, but the system had not been found reliable. There were many prob-

lems involved in the installation of superchargers and ancillary equipment on this Y1B-17A, delaying its first flight until April 1938, three months later than scheduled. Both supercharger and exhausts were fitted above the nacelles and early test flights showed that their operation caused excessive turbulence over the wing, setting up vibration. Efforts to correct this problem came to nothing and eventually Boeing decided that the system would have to be repositioned on the underside of the engine nacelles.

The cost of reworking these installations, nearly $100,000, the Army refused to meet. Perfecting the installation took a considerable number of man-hours and it was not until the early spring of 1939 that all the major problems had been solved. The advantage it bestowed upon the Y1B-17A in performance was considerable. The service ceiling was raised 10,000 ft (3,050 m) more than that of the Y1B-17, and top speed increased by over 40 mph (64 km/hr). The Air Corps, greatly impressed, specified turbo-superchargers in subsequent heavy bomber orders. Meanwhile, later in 1937, Boeing had obtained production orders for ten B-17B models featuring turbo-superchargers and a redesigned nose section; other improvements included a larger rudder, hydraulic brakes, the addition of external bomb racks under the wings, fuel system changes and a reversion to all-metal wing flaps.

Smoothly contoured, the new nose section on the B-17B greatly enhanced the looks of the aircraft in addition to improving its aerodynamic qualities. The restructuring, however, was basically functional. In the Y1B-17 the navigator/bombardier sat behind the pilots and had a restricted outlook. In the B-17B his position was moved to the nose compartment where forward visibility was good and side observation windows gave additional views. By dispensing with the novel, but impractical, revolving nose gun structure of the Y1B-17 and substituting a framed fixture that included a large flat panel, the bombardier's sight could be moved forward. A ball and socket fixture in the Plexiglas framing allowed a 0.30-in (7.62 mm) machine gun to be used for defence. Although the contract for ten B-17Bs was signed in November 1937, followed by orders for an additional 29, technical problems – notably with the turbo-supercharger – delayed production

Y1B-17s over New York City in February 1938. Flown by crews of 2nd BG they are heading for South America on a long-distance flight test. (BOEING)

B-17B with enlarged rudder and larger flaps. (BOEING)

and the first did not fly until late June 1939. Deliveries began in October and were completed in the following March. This order gave the Air Corps sufficient aircraft to fully equip two bombardment groups, one on the Atlantic and one on the Pacific coast of the United States; these were the 2nd Group at Langley Field, Virginia, and the 7th Group at Hamilton Field, California.

With the outbreak of war in Europe, US defence chiefs changed their attitudes towards the development and operation of long-range bombing aircraft and advocates of strategic bombardment propounded their doctrine more openly. Even before delivery of the first B-17B, Boeing had won a contract for 38 improved models, designated B-17C. Moreover, that availability of the B-17B with its supercharged engines allowed investigation and experimentation in high-altitude bombing.

High Altitude Precision Bombing

The Flying Fortress had become the prime instrument in the development of a technique and the promotion of a doctrine. Not only had the Air Corps been furthering the development of turbo-supercharging engines for some years, but also oxygen equipment for crews in the rarefied atmosphere. Another significant development was a high-altitude bomb-sight, (known as the Norden after its inventor) originally procured by the US Navy. Employing gyroscopes and an advanced computing system, it provided a degree of precision far in advance of any other existing bomb-sight. From altitudes of 10,000 ft (3,050 m) bombardiers, using their Norden sights in B-17Bs, could repeatedly place their bombs on or in close proximity to a range target. At higher altitudes the degree of accuracy declined, but it could still be acceptably near to the aiming point even from 20,000 ft (6,100 m) and above.

Thus encouraged, Air Corps planners envisaged these bombers attacking from high altitudes where their speed would make them difficult for contemporary fighters to

reach or pursue, while at the same time they would be above maximum range for most anti-aircraft artillery. Nevertheless, the advance of fighter capability as exhibited in the latest British and German designs and new directing systems, indicated that the B-17B could and would be intercepted. As a defensive measure it would be necessary to fly the bombers in large formations where the massed firepower of their defensive armament would form a formidable deterrent to intercepting fighters.

Although the B-17 had been labelled the Flying Fortress because of its many gun positions, in reality it was poorly armed by European standards. The British – and the French – were busy ordering aircraft from many US manufacturers by the end of the 1930s and were particularly critical of the Boeing's armament. RAF bombers were being fitted with powered turrets and this service considered a tail gun position imperative on large bomber aircraft. The streamlined cupolas on the B-17B – similar to those on the Y1B-17 – proved to have limited field of fire as well as being awkward to operate. The B-17C model was, basically, an attempt by the manufacturers to improve the aircraft's defences. The teardrop-shaped gun cupolas were dispensed with. The two fuselage side waist positions were transformed into open hatches through which

B-17C of 11th Bomb Group. (USAF)

8

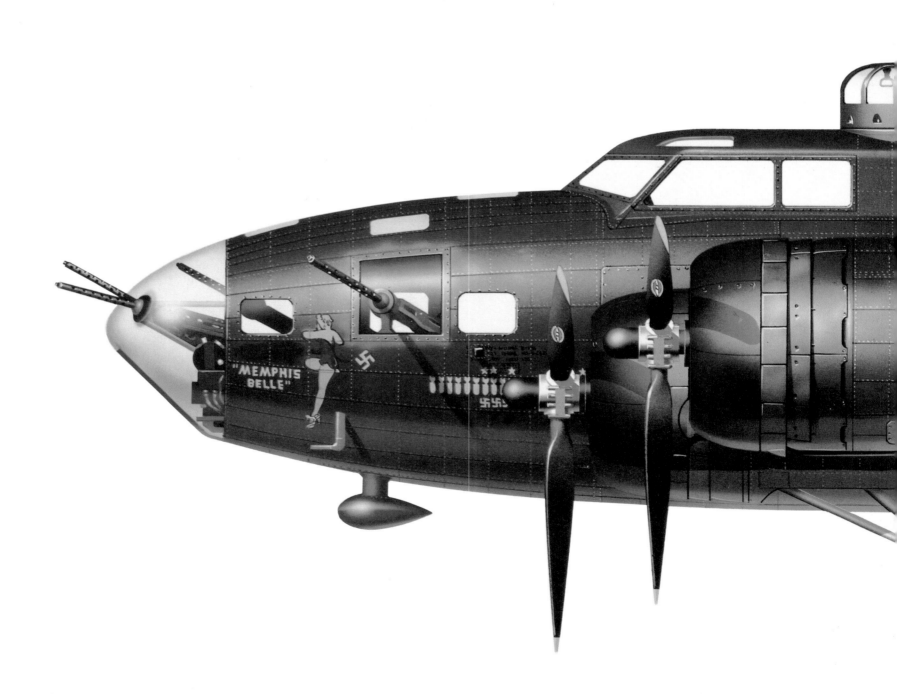

B-17F-10-BO

B-17F "Memphis Belle" of 91st GB, 324th BS, based at Bassingbourn, north of London. Flown by Captain Robert K. Morgan's crew, the Fortress commenced combat operations on 7 November, 1942, the target being Brest in France. "Memphis Belle" completed 25 missions on 17 May, 1943 with an operation against Lorient. The bomber and crew were the first from the 8th AF to be returned to the USA on completion of an operational tour but probably the second to fly 25 missions.

Boeing Model 247D
Modified from Model 247 in 1934. The plane was operated by United Airlines
Power units: 2 × 550 hp Pratt & Whitney Wasp S1H1-G engines
Max speed: 202 mph (325 km/h) at 8,000 ft (2,440 m)
Cruising speed: 189 mph (304 km/h) at 12,000 ft (3,660 m)
Range: 890 mls (1,430 km) at 8,000 ft (2,440 m)
Service ceiling: 25,400 ft (7,740 m)
Span: 74 ft 0 in (22.56 m)
Length, tail up: 51 ft 7 in (15.72 m)
Height, tail down: 12 ft 1.75 in (3.70 m)
Wing area: 836 sq. ft (77.7 m²)
Gross weight: 13,650 lb (6,192 kg)
Crew: 2 pilots and a stewardess
Passengers: 10

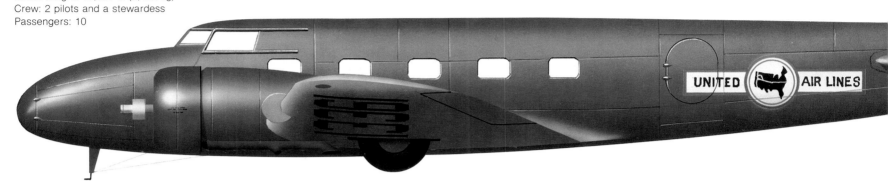

Boeing XB-15
Made its first flight in October, 1937
Power units: 4 × 850 hp Pratt & Whitney R-1830-11 engines
Max speed: 197 mph (317 km/h) at 6,000 ft (1,830 m)
Cruising speed: 171 mph (275 km/h)
Range: 3,400 mls (5,470 km) with 2,511 lb (1,139 kg) bombs
Service ceiling: 18,850 ft (5,750 m)
Span: 149 ft 0 in (45.42 m)
Length, tail up: 87 ft 7 in (26.70 m)
Height, tail down: 18 ft 1 in (5.51 m)
Wing area: 2,780 sq. ft (258.3 m²)
Gross weight: 65,068 lb (29,515 kg)
Armament: 3 × 0.30-in (7.62 mm) and 3 × 0.50-in (12.7 mm) machine guns,
          8,000 lb (3,630 kg) bombs

Crew: 10

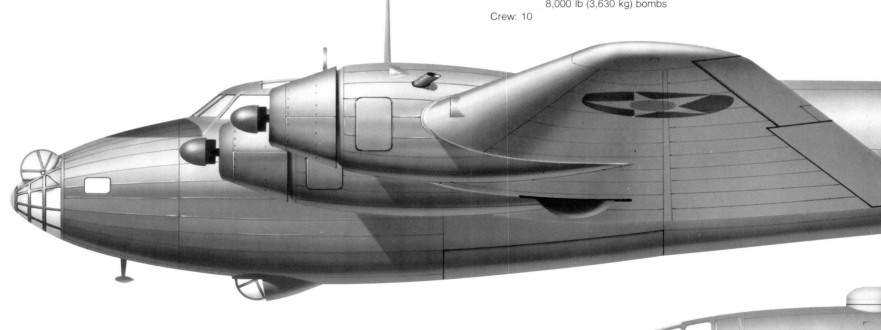

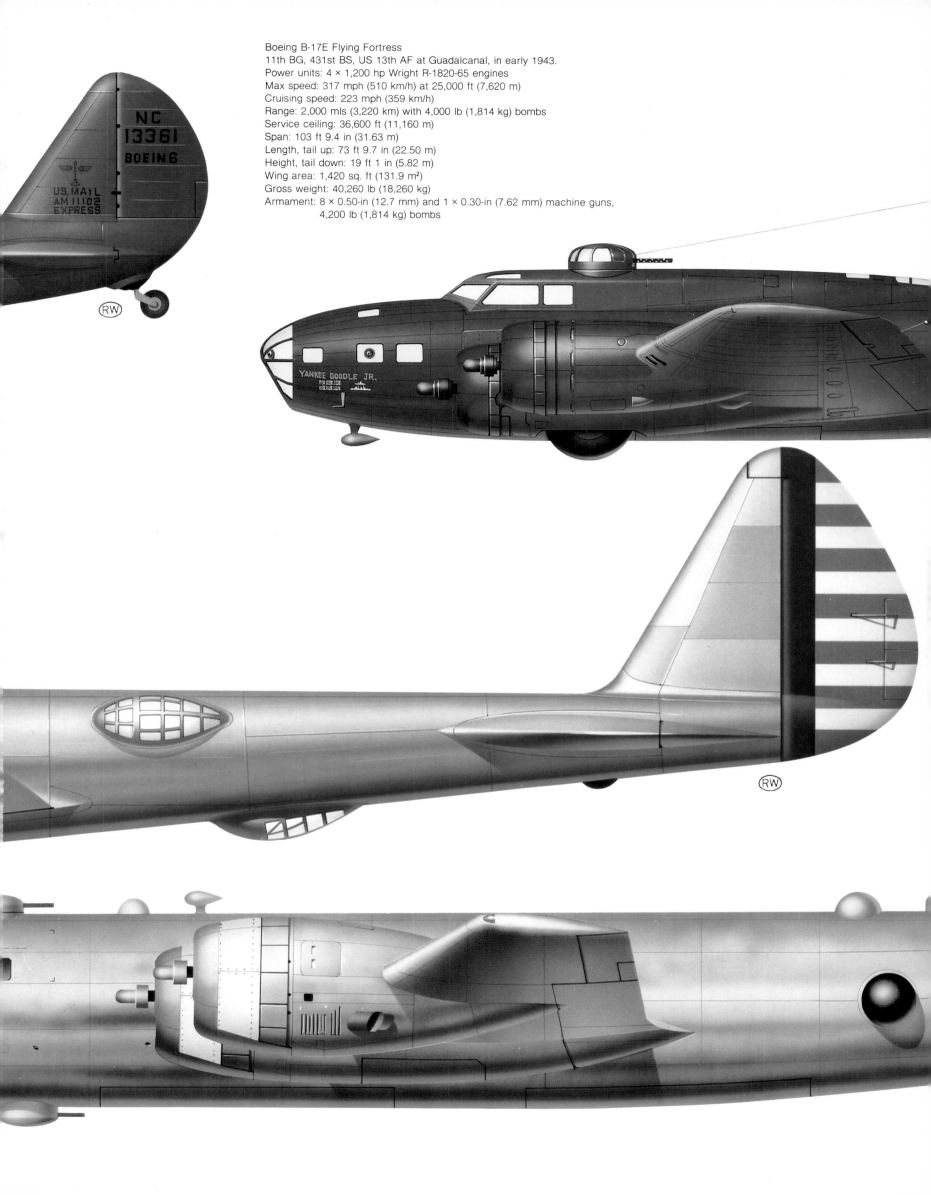

Boeing B-17E Flying Fortress
11th BG, 431st BS, US 13th AF at Guadalcanal, in early 1943.
Power units: 4 × 1,200 hp Wright R-1820-65 engines
Max speed: 317 mph (510 km/h) at 25,000 ft (7,620 m)
Cruising speed: 223 mph (359 km/h)
Range: 2,000 mls (3,220 km) with 4,000 lb (1,814 kg) bombs
Service ceiling: 36,600 ft (11,160 m)
Span: 103 ft 9.4 in (31.63 m)
Length, tail up: 73 ft 9.7 in (22.50 m)
Height, tail down: 19 ft 1 in (5.82 m)
Wing area: 1,420 sq. ft (131.9 m²)
Gross weight: 40,260 lb (18,260 kg)
Armament: 8 × 0.50-in (12.7 mm) and 1 × 0.30-in (7.62 mm) machine guns,
    4,200 lb (1,814 kg) bombs

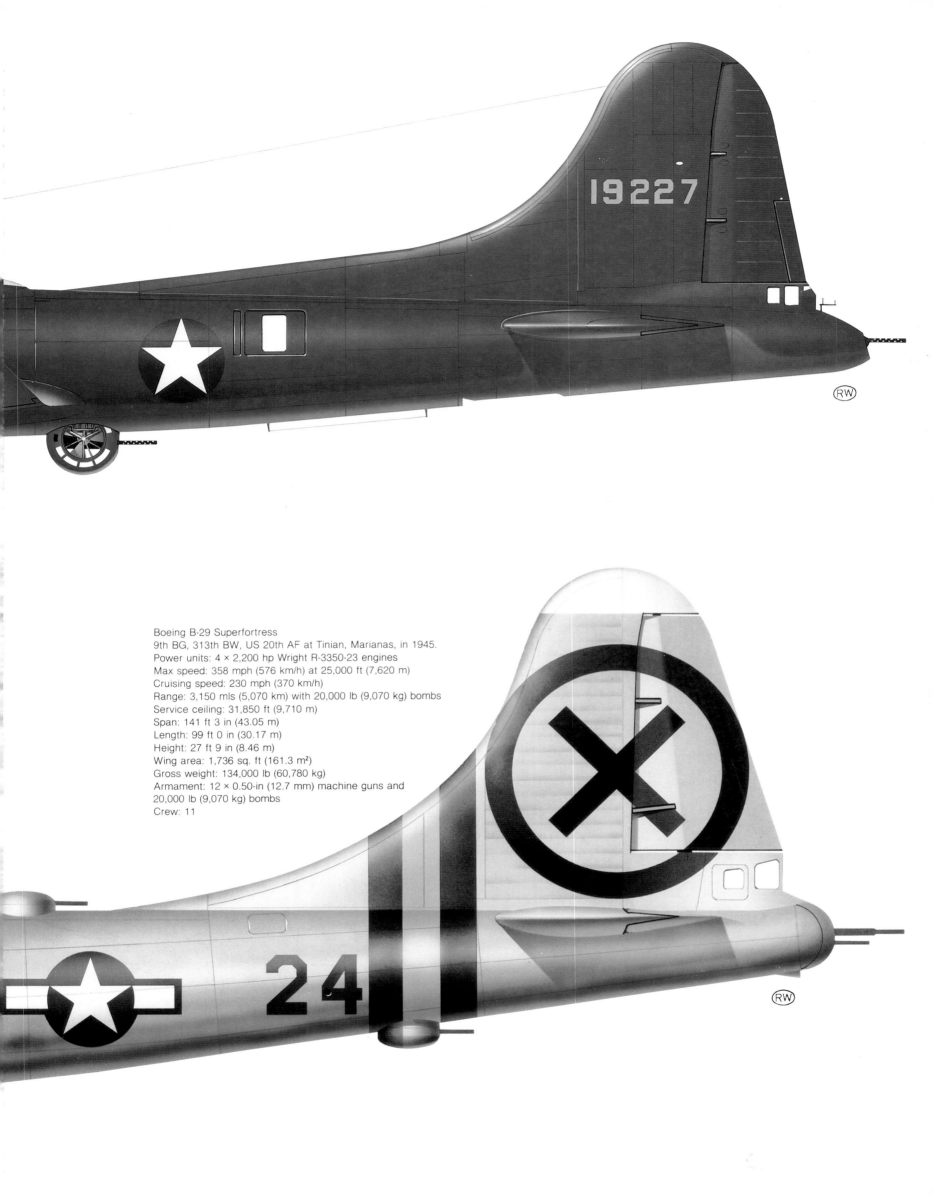

Boeing B-29 Superfortress
9th BG, 313th BW, US 20th AF at Tinian, Marianas, in 1945.
Power units: 4 × 2,200 hp Wright R-3350-23 engines
Max speed: 358 mph (576 km/h) at 25,000 ft (7,620 m)
Cruising speed: 230 mph (370 km/h)
Range: 3,150 mls (5,070 km) with 20,000 lb (9,070 kg) bombs
Service ceiling: 31,850 ft (9,710 m)
Span: 141 ft 3 in (43.05 m)
Length: 99 ft 0 in (30.17 m)
Height: 27 ft 9 in (8.46 m)
Wing area: 1,736 sq. ft (161.3 m²)
Gross weight: 134,000 lb (60,780 kg)
Armament: 12 × 0.50-in (12.7 mm) machine guns and
20,000 lb (9,070 kg) bombs
Crew: 11

a gunner could manoeuvre a single machine gun fixed on a pedestal mount inside the fuselage. In place of the teardrop blister in the roof of the radio room a removable low-profile transparent canopy covered a machine gun hatch. To improve the rearward defence a lowered compartment on the underside of the aircraft, frequently likened to a bathtub, held another machine gun and provided a gunner with a fair field of fire.

The four weapons used to defend the rear and beam of the aircraft were all 0.50-in (12.7 mm) Brownings, whereas frontal defence rested on a lone 0.30-in (7.62 mm) rifle calibre machine gun. The only improvement in this weak area was two additional ball and socket fixtures in the nosepiece as alternative locations for the machine gun. Influenced by developments in Europe, the B-17C also featured some armour plate to protect crew members, and self-sealing fuel tanks. Uprated engines were installed, the R-1820-65 model of the Cyclone rated at 1,200 hp, pushing the top speed to 323 mph at 25,000 ft (520 km/hr at 7,620 m). As with the B model, the C was subject to numerous changes between the singing of the contract and the acceptance of the first aircraft in August 1940. Also, as with the B, the quantity ordered was increased. However, so many changes ensued that the additional 42 aircraft were given a new model number as B-17Ds. Again steps were taken to improve the

Flying Fortress's firepower, this time by installing two 0.50-in (12.7 mm) Browning machine guns in the upper and lower gun positions.

Carrying heavy loads to high altitude brought high engine temperatures and, in an effort to improve cooling, trailing edge flaps were fitted to the engine cowlings of the D model. In fact, this was the only major external difference between the C and D and although most Cs and many Bs went for modification and received many D features, these apparently did not include the cowling flaps. The D also had revised electrical, fuel and oxygen systems, extra armour plate, bomb rack changes and improved self-sealing fuel tanks – the original type having been prone to leaks!

From a half-dozen US companies working on power turret designs Boeing planned to install those being produced by Sperry on late production B-17Ds. But as development problems delayed the production of these turrets, only one experimental installation was carried out, the turret being positioned just aft of the cockpit where the gunnery controller's Plexiglas dome was sited on the B-17C and D. The majority of B-17Ds were delivered to the Air Corps during the early spring of 1941, most being sent to reinforce the 5th Bomb Group in Hawaii and to re-equip the 19th Bomb Group which moved from the US to the Philippine Islands later in the year.

B-17C under flight test. (BOEING)

B-17D with cowl flaps that were introduced with this model. (BOEING)

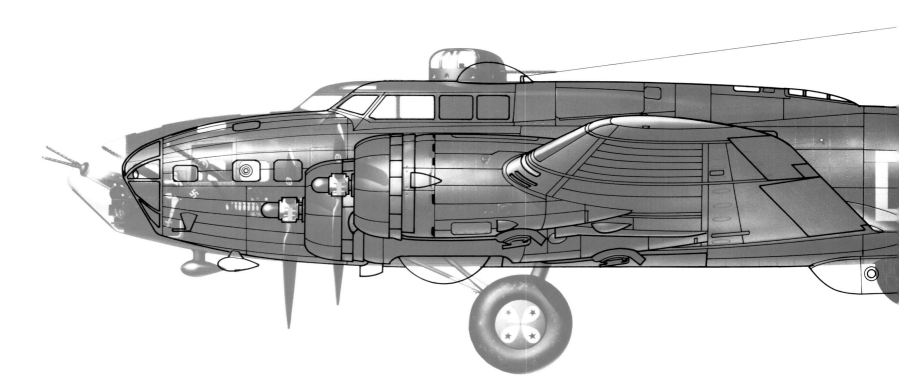

First Combat

The first Flying Fortresses used in war operations were flown by Royal Air Force crews. This acquisition of the B-17s by the British is in itself interesting and although the episode can hardly be said to have been successful, it provided some important information on the operation of the aircraft and high-altitude bombing over Europe.

When, in the summer of 1940, Great Britain stood alone against the Axis powers, she was anxious to acquire as much war material as possible from the United States. High on the shopping list were aircraft, and large quantities were supplied at first by direct purchase and then under Lend-Lease. The B-17 was not a priority as, prior to hostilities, RAF bomber leaders had inspected the aircraft and considered both its bomb-carrying capacity and defensive armament inadequate. Nevertheless, they knew the aircraft to be well made and to have good handling qualities. It appeared to them to be ideal for maritime-reconnaissance in view of the growing U-boat menace.

The US government, eager that the public should see some return for their money being lavished on armaments, particularly aircraft, thought it would be useful if the pride of the Air Corps could exhibit some of its prowess and gain some publicity on the quality of their goods. The British government, on the other hand, was interested in any move

that would appear to align the United States with their war aims. The propaganda value of being able to launch the vaunted American Flying Fortress against Germany was considerable. The US and British governments came to a decision whereby 20 B 17Cs revamped to D standard would be sent to the United Kingdom and an RAF squadron be specially trained to operate them according to the method of high-altitude precision bombing developed by the Air Corps. Although reluctant to part with their precious B-17s, particularly to a service known to be sceptical of their use in bombing, the US Army Air Corps (USAAC) – which became the US Army Air Force (USAAF) in April 1941 – was anxious to see that the aircraft had a good press. Although the United States was officially still neutral, arrangements were made for a small number of experienced pilots and technicians to go to England to train and advise RAF personnel on the use of the bomber.

In RAF service the B-17 was known as the Fortress, as it was British policy to use names rather than designations to identify their aircraft types. The RAF, like the Luftwaffe, had tried daylight bombing operations with large aircraft and found them extremely vulnerable to interception. In consequence RAF heavy bomber operations were conducted under cover of darkness and only fast low-level attacks by light bombers were carried out in daylight. From their experience, RAF Bomber Command considered that the only way the Fortresses could be employed in high-altitude day bombing

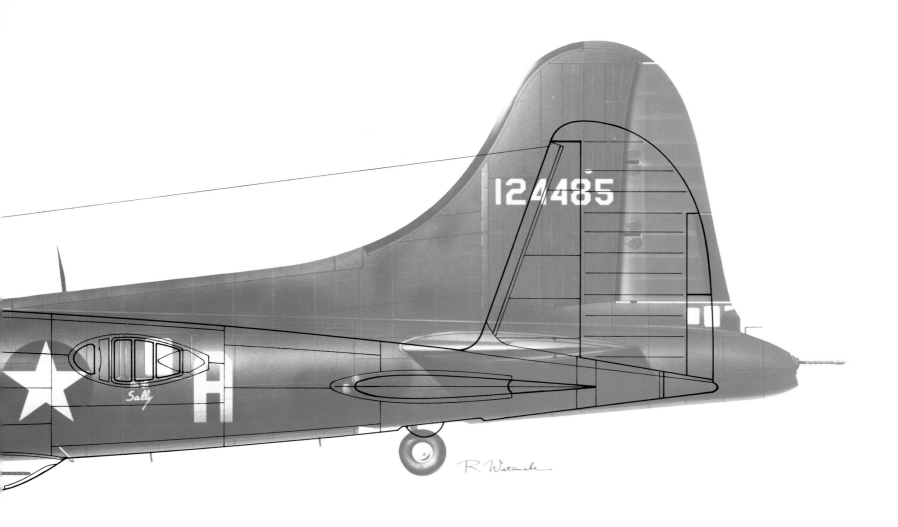

Fortress I of No. 90 Squadron, RAF. This aircraft was involved in the air fight on 2 August, 1941.

raids was to make maximum use of their then very high ceiling. The plan was to operate the Fortresses at an optimum 32,000 ft (9,750 m), a height which the known types of enemy fighter would have difficulty in attaining. This was, however, some 10,000 ft (3,050 m) higher than the Air Corps had planned in developing their technique of high-altitude bombing; in the event, this extra altitude proved critical. The still secret Norden bomb-sight had not been released to the British and a Sperry sight was substituted.

RAF crews specially selected and trained were allotted to 90 Sqn re-formed for this task on 7 May, 1941 at RAF Watton in Norfolk. Their first operation was not flown until 8 July due to various problems with the aircraft, airfield moves and awaiting suitable weather – a clear day was essential for target location and bombing. Three Fortresses were despatched from Polebrook in Northamptonshire on the first operation with the Wilhelmshaven naval base as the objective. Only one bomber was able to complete the mission as planned; the other two were plagued by technical problems caused by the high humidity of the atmosphere and the extremely low temperatures. Windshields and Plexiglas frosted thus obstructing vision, bomb release and gun mechanisms were also frozen and a supercharger failed. The engines of one Fortress began to throw oil from the crankcase breathers after the bomber had reached 23,000 ft (7,010 m) and, continuing to do so, forced the pilot to abandon his mission. Oil throwing remained a major problem but, strangely, only certain engines were affected. From problems encountered on this and subsequent operations it became evident that the Fortress could only rarely be operated at the desired altitude of 32,000 ft (9,750 m). Consequently Fortresses were used mainly singly and often had to attack their targets from lower altitudes with the attendant risk of interception.

The first brush with the enemy occurred on 2 August, 1941 when Pilot Officer F. W. Sturmey's crew flying Fortress AN519 were attacked off the Dutch coast at 22,000 ft (6,710 m) by two Bf 109s. The result of the exchange of fire was in the Fortress crew's favour as, unknown to them, one Messerschmitt receiving hits in the engine was abandoned by the pilot. The other Bf 109 was also hit and the pilot had to make an emergency landing in Holland. This incident proved the exception for, in following actions between RAF Fortresses and enemy fighters, it was the bombers which suffered. Two weeks after this successful engagement Pilot Officer Sturmey was flying another Fortress near Brest, when it was repeatedly attacked and severely damaged. Two of the gunners were killed but Sturmey managed to bring the aircraft back for a crash-landing in England.

On a mission to Oslo on 8 September, 1941, a Fortress was shot down, the first of its kind to fall to enemy action. Another Fortress also failed to return from this operation although it was believed to be lost through an accident. At the end of the month the squadron was taken off operations. Bombing accuracy had been poor due to the physical and mental strain imposed using oxygen and heating equipment. Although the armament was found inadequate, no enemy fighters had been able to intercept when the aircraft

Wright R-1820 nine-cylinder air-cooled radial engine.
(CURTISS-WRIGHT CORPORATION via BRIAN D. O'NEILL)

were flying above 32,000 ft (9,750 m). Had the aircraft been able to maintain altitudes above this level it might have been able to operate with impunity. The RAF was also critical of the average bomb-load which they considered uneconomical in relation to the manning and maintenance effort.

Despite the lack of success some favourable propaganda was produced for this effort by both British and US news media. A great deal of useful information had been gathered which was put to good use by both the USAAF and Boeing. Perhaps not enough attention was paid to the problems of operating in the moisture-laden air of Europe, for difficulties encountered by the RAF in 1941 were still to be solved when the USAAF brought the Fortress to Europe again late the following year.

Revamping Design

While RAF experience had again underlined the poor defensive firepower of the B-17, Boeing had already taken some positive steps to really transform the B-17 into a flying fortress. Advances in military aviation technology were so rapid during the late 1930s that, by the time hostilities had commenced in Europe, Boeing were aware that the advanced bomber design they had created in 1934 was fast approaching obsolescence. Another US manufacturer, Consolidated, was producing a new four-engined bomber, the B-24, which would have an all-round performance superior to the B-17. It had also been designed to take a power-operated gun turret in the tail, the most vital defence point on any large bomber. With the possibility that the United States herself might soon be involved in the war, it seemed likely that funds for a substantial order for an improved B-17 type would be forthcoming.

The immediate problem was whether the B-17 could be developed further to meet the defensive requirements that were bound to be imposed by the Air Corps or if it would involve a completely new design. Time was a critical factor and Boeing decided to revamp the B-17 by designing a completely new rear fuselage in which the tail gun position could be incorporated. At the same time a larger tailplane was developed to overcome complaints of lateral instability at high altitudes. These and other changes proposed to the Air Corps in the summer of 1940 were duly accepted, resulting in the new B-17E being put into production early the following year. As everything aft of the radio compartment was new, it is surprising that the Air Corps did not bestow a completely new model designation. Certainly the change gave the Fortress a completely different appearance for the shape of the empennage bore no resemblance to the earlier models. The outstanding visual feature was the large fin which was derived from that on the Boeing Model 307 Stratoliner, the commercial airliner developed in conjunction with, and using the same basic wing and powerplants as, the B-17.

Although a major reason for the redesign was the need for a tail gun position, Boeing did not install one of the new power-operated turrets. Being bulky, their inclusion would have involved a larger diameter rear fuselage or a bulge behind the tail. To avoid increasing drag Boeing preferred to maintain the tapering fuselage, terminating in a small gunner's compartment where twin 0.50-in (12.7 mm) guns would be hand-operated by a gunner in a sit-kneel position. The two beam or waist gun hatches were retained but were now in rectangular form. The entrance door, previously placed near the starboard wing root, was repositioned on that side between the tailplane and the waist gun hatch. For ventral defence a compact two-gun remotely controlled under-turret was located in the forward part of the new

fuselage section. It was sighted by a gunner in prone position using a periscope device projecting through the bottom of the fuselage to the rear of the turret – a method that proved far from satisfactory.

This Bendix turret was really a stopgap until the new manned Sperry under-turret became available. A Sperry electrically powered upper-turret was installed aft of the cockpit in the position worked out for the B-17D. Nose armament remained unchanged as frontal interception of such a fast bomber was considered unlikely to be attempted.

The B-17E, however, was less speedy than its predecessor, as the larger tailplane and the drag induced by the turrets reduced top speed by some 6 mph to 317 mph at 25,000 ft (510 km/hr at 7,620 m). The revised fuselage increased the length of the aircraft from 68 ft 4 in to 73 ft 10 in (20.8 m to 22.5 m), and the height at rest from 15 ft 5 in to 19 ft 2 in (4.7 m to 5.84 m). Another feature incorporated was a fully retractable tailwheel. Certainly in the B-17E Boeing had achieved an improvement in both firepower and field of fire (there were now eight 0.50-in [12.7 mm] guns) with minimal effect on the aerodynamic qualities of the airframe. Indeed, they had produced a classic configuration with undeniable aesthetic appeal.

The first B-17E did not fly until 5 September, 1941, more than four months behind schedule. The problem then was shortage of materials. Faced with the prospect of involvement in hostilities the US government had placed huge orders for a vast range of military equipment, temporarily swamping America's industrial capacity. Orders for 512 B-17Es had been obtained in the late summer of 1940, and to boost production still further agreements were established with Douglas and Lockheed-Vega for additional production of the B-17E in their plants. Boeing themselves were now trying to expand as rapidly as possible, building new factories and expanding old. A completely new bomber design was in the pipeline, the B-29, but if the US became involved in a war at an early date the B-17 and Consolidated's B-24 would be the only heavy bombers ready for operation.

A new B-17E at Boeing's Seattle plant on 19 February, 1942.

(BOEING)

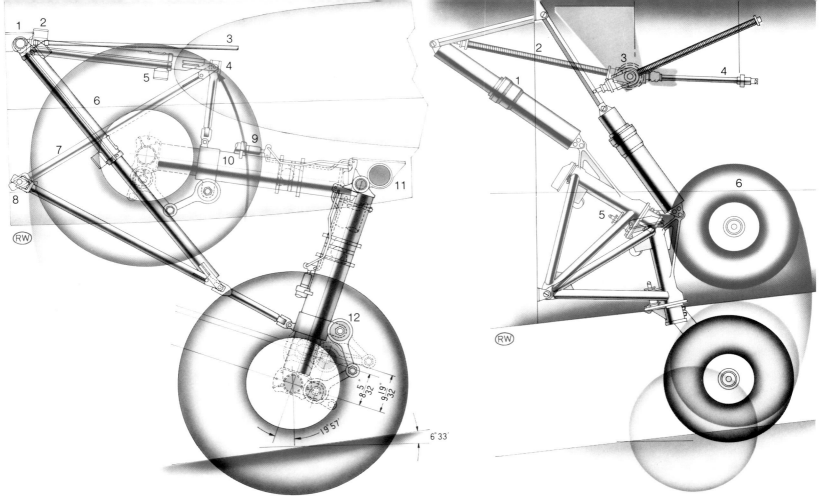

Main wheel diameter: 55 in (1,397 mm)

Landing gear retracting mechanism
1 Retracting screw support
2 Upper limit switch
3 Control installed hand retraction
4 Drag strut universal assembly
5 Lower limit switch
6 Retracted tyre
7 Drag strut
8 Drag strut support
9 Return booster valve
10 Strut assembly
11 Landing gear support
12 Torsion link assembly

Tail wheel diameter: 26 in (660 mm)

Tail gear retracting mechanism
1 Oleo
2 Retracting gear
3 Tail wheel motor
4 Hand retracting mechanism
5 Tail sheel lock assembly
6 Retracted tyre

In Action Against the Japanese

B-17s in fact became involved at the very opening of hostilities between Japan and the United States when a small number, sent from the US to Hawaii as reinforcements, were caught up in the Japanese attack on Pearl Harbor. In the following days 53 B-17Es were despatched to the Philippines, Java, Australia and Hawaii to help combat the Japanese invasion through the Central and South-east Pacific. During the initial attacks on bases in the Philippines most of the 35 B-17C and D models assigned to the 7th and 19th Bombardment Groups were destroyed by strafing Japanese fighters. Surviving aircraft moved to Java and Australia, to be joined by the reinforcements from the United States for flying maritime-reconnaissance and occasional bombing attacks on enemy shipping. With a range of 2,000 miles (3,220 km), the B-17 was one of the few aircraft available that could be used to reconnoitre the vast stretches of ocean and report on Japanese movements. But the small number of aircraft available, lack of parts and poor maintenance facilities limited the average strike force to only three or four and there was little effective bombing.

Nevertheless, the Flying Fortress was able to give quite a good account of itself and live up to its name on occasions. With one exception, current Japanese fighters were only armed with two rifle calibre machine guns with unsuitable ammunition including rounds with a 6.35 mm (0.25-in) armour-piercing core. It proved extremely difficult

for Japanese fighter pilots, in even the most advantageous circumstances, to bring down a B-17C or D. Their armour-piercing ammunition was not effective against the Boeing's armour and though riddled with bullet holes, B-17s usually survived combat with Japanese fighters. The famed Zero, the Japanese Navy Type 0, was at that time the only antagonist equipped with 20 mm cannon capable of inflicting serious damage. However, B-17s could survive Zero attacks due to the fact that the lack of 20 mm shell fuzing resulted in impact explosion causing only superficial damage. At this stage of the war Japanese fighters lacked adequate armour protection and self-sealing fuel tanks and so were particularly vulnerable to accurate defensive fire. The B-17E's tail guns proved particularly useful in warding off attacks.

Japanese pilots developed a healthy respect for the B-17 and as a result of these early encounters improvements were made in the firepower of their fighters. Subsequently it became regular practice to make frontal attacks on B-17s where it was known their defensive armament was poor. Combats in the South-west Pacific War theatre also revealed that the early B-17E's remotely controlled under-turret was ineffectual, leading to its removal. Hand-held guns were often mounted to fire through the open space where the turret had been located. With the 113th B-17E in production, the manned Sperry under-turret was introduced. Because of its unique spherical structure, it was termed a "ball turret". Unlike most powered turrets where the guns were elevated or depressed manually, those in the Sperry ball were fixed and it

B-17Es shortly after delivery to USAAF.

was the whole turret that moved. The gunner sat in a foetus-like position between the guns, using a reflector sight aimed between his feet. His back was to the entry door through which access could be gained on the ground.

In practice a gunner would not be in position during landing or take-off in case of undercarriage failure. For entry, usually made once the bomber was airborne, it was necessary to revolve the turret so that the guns pointed downward to expose the door. The whole electrically operated assembly was supported on a yoke and gimbal which, while looking rather unsubstantial, was actually very robust and in a crash would tear the fuselage rather than break the attachments. Even so, the ball turret was not a place for the claustrophobic; it was extremely compact and men of small size were usually picked for this position. Owing to space restriction a parachute could not normally be worn and in an emergency the gunner had to go back into the fuselage to get his.

The ball turret proved highly successful and was of great value in fighting off attacks from below. It did, however, bring one problem that was never satisfactorily resolved: cases from the two 0.50-in (12.7 mm) guns were ejected outside the turret and on occasions caused damage to other Fortresses in the formation.

A total of 812 B-17Es were ordered from Boeing, but only 512 were completed as such as the rest came under a new designation, B-17F. The F can be described as the first really battle-worthy Fortress – this name had also been officially accepted by the USAAF. Apart from the nosepiece devoid of framing, the F was externally similar to the E model. Internally, it was subject to some 400 changes, albeit most of them minor. One of the most important was the Wright R-1820-97 model of the Cyclone turning Hamilton Standard "paddle blade" propellers which had a diameter of 11 ft 7 in (3.53 m), an inch more than the type previously fitted. The changes necessitated the leading edge contours of engine cowlings being remodelled to avoid the wider blades striking them when fully feathered. Other improvements were to the landing gear, brakes, and the oxygen system. There were also changes in the bomb racks, ball turret, and the addition of an automatic pilot/bomb sight link. While the new model engines developed the same power as those in the B-17E, and were subject to only detail changes such as carburettor dust filters, the new wide blade propellers gave improved performance at high altitude, raising the service ceiling to 38,000 ft (11,580 m) and increasing top speed to well over 320 mph (515 km/hr). During the production run another 10,000 lb (4,536 kg) weight was added through various equipment improvements which in turn adversely affected performance, reducing the top speed to 299 mph (481 km/hr).

The Wright R-1820 Cyclone was near the limit of its

XB-38

Modified from B-17E

Power plants
 4 × Allison V-1710-89 liquid cooled engines:
 1,425 hp at 25,000 ft (7,620 m)

Performance
 Max speed: 327 mph (526 km/h)/25,000 ft (7,620 m)
 Cruising speed: 226 mph (364 km/h)
 Service ceiling: 29,700 ft (9,050 m)
 Max range: 3,600 mls (5,790 km)
 Range with 6,000 lb (2,720 kg) bombs: 1,900 mls (3,060 km)

Weights
 Empty: 34,748 lb (15,762 kg)
 Gross: 56,000 lb (25,400 kg)
 Max take-off: 64,000 lb (29,030 kg)

Dimensions
 Same as B-17E

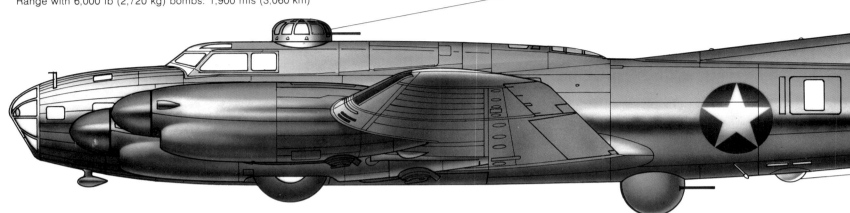

development and further improvements in the Fortress's performance were only likely to be achieved by installing a completely new engine type. The liquid-cooled supercharged Allison V-1710-89 engine rated at 1,425 hp offered a potential improvement in performance, although the Army Air Forces had long favoured air-cooled radials for their reliability and simpler maintenance. Lockheed-Vega were asked to carry out an experimental installation of Allisons in a Fortress airframe to be designated XB-38, in line with the USAAF policy of bestowing a new aircraft model number when major changes in powerplants were made. A B-17E previously obtained by Lockheed for familiarization purposes when setting up their Fortress production line at Burbank, California, was used for the conversion.

Work began in the summer of 1942 and progressed slowly and it was not until 19 May, 1943 that the XB-38 took to the air. It performed well in early tests, but had to be grounded for some days when exhaust manifold leaks were discovered. During a subsequent test, four weeks after its first flight, the XB-38 had to be abandoned by its crew because of an uncontrollable engine fire, the co-pilot being killed when his parachute failed to open; the aircraft crashed in open country. The data obtained from test flights attributed a performance comparable to that of the B-17F at similar weight. Apparently there had been no opportunity to find the service ceiling that the turbo-supercharged Allisons would give the XB-38, a calculated figure being substantially less than that of a Cyclone-powered Fortress. Plans to build two more B-38s, and the actual conversion of the Lockheed-Vega Fortress production line to the type, were dropped when it was decided that its extra weight would limit endurance and offer few, if any, advantages over the radial-engined types.

Deliveries of B-17Fs began in late May 1942 from the Boeing plant, followed by the first from the Douglas line at Long Beach in California, and later in the summer from

Lockheed-Vega. The two new factories took time to build up to a mass production schedule and by the end of the year Douglas had only delivered 85 aircraft and the Vega plant 68, whereas Seattle turned out 850. By the time the F was superseded by the G model in the late summer of 1943 the combined production from all three plants was averaging 400 aircraft a month.

Large-scale service use of the Fortress highlighted the need for many improvements or modifications to the aircraft and its equipment. Changes could not usually be introduced quickly on the production line without some disruption and as the priority was to build increasing numbers of aircraft, production flow took first place. To effect the desired changes speedily, modification centres were set up to receive aircraft after they left the factory. This policy was continued with much success until the end of hostilities, while other modification centres overseas incorporated changes desirable in particular theatres of war.

A theatre modification carried out to many B-17Es and most Fs was the installation of additional defensive firepower in the nose of the aircraft. Japanese frontal attack tactics had resulted in many hastily improvised nose gun installation in B-17Es operating in the South-west and Central Pacific battle zones. These varied from a twin-gun mount firing through an opening in the top of the Plexiglas nose, to single 0.50-in (12.7 mm) weapons in openings made by removal of the side window Plexiglas. In the light of this requirement, modification centres in the US installed a 0.50-in gun and mount to fire through an aperture made in Plexiglas observation windows on both sides of the nose. These were staggered to allow more freedom of operation in the compartment and could be used by either the bombardier or the navigator, although in practice it was usually the navigator who manned these guns. Larger gun windows later became standard, allowing a wider field of fire forward.

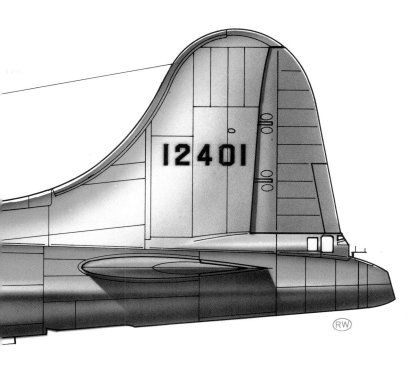

The Main Air Front

In July 1942 the USAAF began moving B-17 units to the United Kingdom with the intention of building a strategic bombing force for operations against Germany. The first unit to arrive, the 97th Bombardment Group equipped with B-17Es, commenced operations over France on 17 August. During that month other groups arrived with early B-17Fs and a few weeks later the E models were relegated to operational training duties. The two most experienced Fortress groups were sent down to French North Africa following the Allied invasion of November 1942. This left four B-17 groups and a small number of B-24s in England to conduct strategic bombing missions during the winter and following spring. Initially the high-flying bombers, usually in group formations of 18 to 21 aircraft, gave a good account of themselves against enemy fighter attacks. Bombing results, varied at first, improved when experienced crews led and sighted on target, with the rest of the formation briefed to follow their signal for bomb release. Luftwaffe fighter tactics then changed and included frontal attacks on the B-17s.

As in the Pacific, defensive armament improvements were carried out locally with one or two 0.50-in (12.7 mm) machine guns on special mounts firing through an aperture in the Plexiglas nosepiece. The 0.30-in (7.62 mm) machine gun was abandoned completely, although it continued to be specified for production B-17Fs for some time. With the field additions B-17Fs operating from England were carrying 12 or 13 0.50-in machine guns, making the aircraft the most heavily armed bomber in service with any of the belligerents. The authorized crew of a B-17F was ten men, the official change having been recorded with the introduction of the "big tail" Fortresses. Early models had six-man crews, increased to seven or eight when hostilities commenced. Of the ten-man crew, four in the forward part of the aircraft were officers – the pilot (who sat in the left-hand seat), the co-pilot and the navigator and bombardier in the nose section. The remaining men were all non-commissioned officers. The flight engineer monitored engine operation and fuel supply amongst other duties and also manned the top turret guns. The radio operator also doubled as a gunner using the machine gun in the roof of his compartment. The remaining four men were gunners manning the tail and ball turrets and the two waist positions. Fortresses used for leading formations often carried one or two additional men for special purposes. In some theatres of war only one waist gunner was carried, reducing the average crew number to nine.

In early 1943 the USAAF operating agency in the United Kingdom, the 8th Air Force, extended its bombing missions into Germany. On some operations losses were high but the Fortress performed so well, both in fighting off enemy interceptors and returning safely to base after sustaining heavy battle damage, that the USAAF were encouraged to commit larger forces to these unescorted high-altitude daylight bombing missions. In May 1943 the B-17 force in the UK was doubled and by August a total of 16 B-17F groups were in action. Two other B-17F groups scheduled for the 8th Air Force had been diverted to North Africa earlier in the year to give a total of 20 groups fully equipped with this model in action against Germany and Italy.

Douglas Long Beach plant built B-17F-30-DL of 381st BG, 535th BS, US 8th AF flying over English countryside in 1943. (USAF)

XB-40 armed with 14 Browning 0.50-in (12.7 mm) machine guns, which reduced the maximum speed to 292 mph (470 km-h) at 25,000 ft (7,620 m).

Among new arrivals in England in the spring of 1943 was a squadron equipped with special Fortresses designated YB-40. These aircraft had extra heavy armament and armour for acting in an escort role. The intention was that they would draw the attention of intercepting fighters and afford the B-17 bomber formation they were covering some protection. A prototype, the XB-40, was converted by Lookheed-Vega from the second Boeing-built B-17F. Firepower was increased by installing twin machine guns in each waist window position, a second two-gun upper turret in the radio compartment and a remotely controlled two-gun turret in the chin position under the nose for forward defence. The chin turret, developed by Bendix and based on the earlier ventrally situated remote-control turret on some B-17Es, was operated and sighted by a gunner sitting in the bombardier's position above.

Altogether, the XB-40 had 14 0.50-in (12.7 mm) machine guns, only one more than many B-17Fs already operating from England. However, the additional power turrets plus power boost to the manipulation of both tail and waist guns, gave the XB-40 the potential of being far more effective against enemy fighters than the B-17. To feed the battery of guns more than 11,000 rounds were carried, approximately three times that specified for a B-17F. In fact, the XB-40's ammunition load could be boosted to over 17,000 rounds by utilizing the space available in the bomb-bay, as it was not intended that the XB-40 should be used for bombing.

Approved by the USAAF during the winter of 1942–43, the prototype XB-40 was followed by an order for 12 YB-40s to be used for an operational test from England. The conversions were carried out on Lockheed-Vega production B-17Fs, the modifications and equipment being similar to those on the XB-40. However, the YB-40's armament was further increased to 16 guns by the addition of two nose side guns, as the standard modification fixture on all B-17Fs. YB-40s were assigned to the 327th Bombardment Squadron at

YB-40

Modified from B-17F-10-VE
Armed with 16 × 0.50-in (12.7 mm) machine guns. Assigned to 92nd BG, 327th BS of US 8th AF based at Alconbury in England, June 1943.

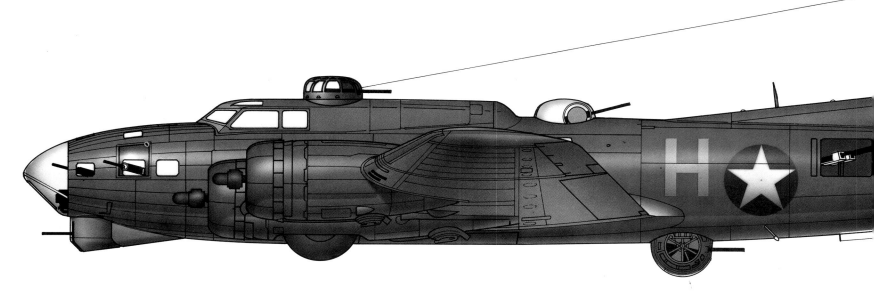

Alconbury, England, and they were despatched on a number of missions between late May and late July.

The aircraft was not a success primarily because of weight distribution; its flying characteristics were poor because of a shift in the centre of gravity towards the rear of the aircraft which made control at high altitudes and formation flying difficult. Moreover, once the B-17s had dropped their loads, the YB-40s were unable to keep up with them, as they were still laden with ammunition and were already a ton heavier than a B-17F empty. With such flight problems the YB-40s could afford little help to a Fortress formation under attack – indeed, it appeared to be more of a liability than an asset as Fortress formations sometimes had to reduce power to stay with them. Meanwhile a further 11 YB-40s had been ordered, but as far as is known only seven were completed before the USAAF decided to abandon the type.

While modification centres made many changes to Fortresses before they reached combat units, changes were constantly effected on the production line. To ease identification of variations in a model, a system of production block numbers had been introduced in 1942 whereby most changes made on a production line were identified. As the three B-17 factories did not always produce identical models with the same equipment at the same time, block numbers were further identified by code letters distinguishing the manufacturer. The B-17F received 26 different block changes at Seattle, 17 at Long Beach and 11 at Burbank. The majority of changes involved internal equipment, usually the substitution of improved types.

The most important and major improvement to the B-17F model came with the B-17F-80-BO, B-17F-25-DL and B-17F-30-VE when two additional fuel tanks of 270 US gal (1,022 ltr) capacity each were placed in each outer wing. This raised the total built-in tankage of the B-17 from 1,730 US gal capacity to 2,810 (6,550 to 10,636 ltr), increasing the stated range from 1,300 miles to 2,200 (2,092 to 3,540 km) at

comparative loadings and settings. In the high altitude operations conducted over Europe the additional fuel gave an extra 3 hr endurance. However, the increased fuel load, by adding another 6,360 lb (2,885 kg) to the B-17F's gross weight, reduced the rate of climb, so that it took 38 min to reach 20,000 ft (6,100 m) compared with 25 min in the earlier B-17Fs. The additional tanks became popularly known as "Tokyo Tanks" because it was said that they would enable a B-17 to make a one-way trip to Tokyo. Although bestowing increased range the Tokyo Tanks came to be seen as something of a mixed blessing. Fuel fumes collecting in the outer wing sections were vulnerable to incendiary ammunition by explosion and fire hazarding the bomber. It was some time before a system of natural air venting was installed in Fortress wingtips to reduce this risk.

As previously stated, the B-17F could absorb a considerable amount of battle damage and survive. The flying qualities of the aircraft were so good that it could lose one side of a tailplane or considerable areas of fin and rudder and still return to make a safe landing. Aircraft often returned with one or two engines out of action and sometimes with a main fuel tank burned out. Damage to hydraulic systems forced a number of wheels-up or "belly" landings and provided the ball turret was jettisoned and the tailwheel lowered, the damage to the airframe in landing could be slight and soon repaired. If the ball turret remained in place and the tailwheel was not extended, the turret suspension frame was usually forced through the base of the fuselage causing it to distort. In such instances economic repair was not possible and the aircraft was written off.

Service Troubles

The Fortress was a fairly viceless aircraft, but operation in the rarefied air of the sub-stratosphere put considerable strain on both crew and aircraft. Engine temperatures ran to near critical limits and had to be constantly monitored. The sub-zero temperatures caused equipment to seize up with frost, particularly in the early days of the offensive before the art of keeping moisture from gun and bomb rack mechanisms was more practised. A major problem was freezing of the hydraulically actuated turbo-supercharger regulators. To prevent this they had to be worked at frequent intervals. Sluggish operation of the turbos was a frequent complaint from pilots. Another problem encountered, whether caused by enemy action or mechanical failure, was the inability to feather a propeller. It was found that when an engine was damaged or failed, loss of fluid would not allow hydraulic adjustment of the propeller blades to a position where the propeller would remain stationary and not windmill. In consequence idling propellers could pick up a high turning rate causing such severe vibration that the safety of the aircraft was in jeopardy. A number of aircraft were lost both directly and indirectly to this cause; it was some months before the extent of this trouble was realized and modifications were introduced.

Crew discomfort was considerable: the heavy clothing needed to keep out the cold impeded movement about

US general purpose bomb

Total weight: 500 lb (227 kg)
Explosive weight: 255 lb (116 kg)
Length overall: 4 ft 11⅕ in (150.3 cm)
Fin width: 1 ft 7 in (48.1 cm)

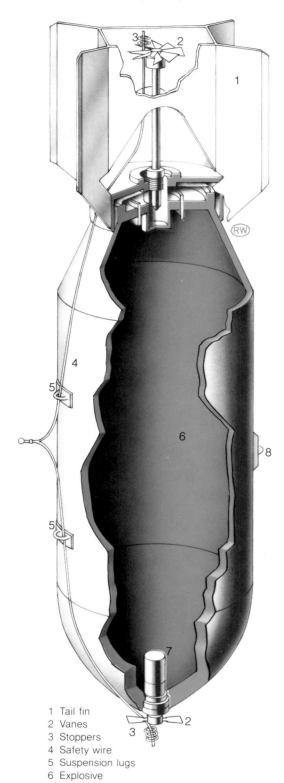

1 Tail fin
2 Vanes
3 Stoppers
4 Safety wire
5 Suspension lugs
6 Explosive
7 Exploder container
8 Lug

US semi-armour piercing bomb

Total weight: 1,000 lb (454 kg)
Explosive weight: 310 lb (141 kg)
Length overall: 5 ft 10⅖ in (178.8 cm)
Fin width: 1 ft 8⁷⁄₁₀ in (52.6 cm)

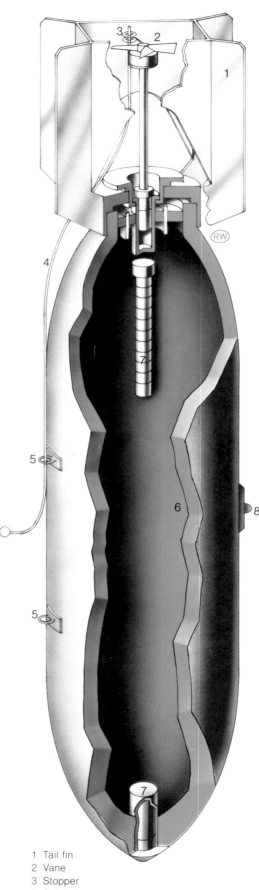

1 Tail fin
2 Vane
3 Stopper
4 Safety lug
5 Suspension lugs
6 Explosive
7 Exploder containers
8 Lug

US armour piercing bomb

Total weight: 1,000 lb (454 kg)
Explosive weight: 145 lb (66 kg)
Length overall: 6 ft 1 in (185.4 cm)
Fin width: 1 ft 4⅔ in (42.2 cm)

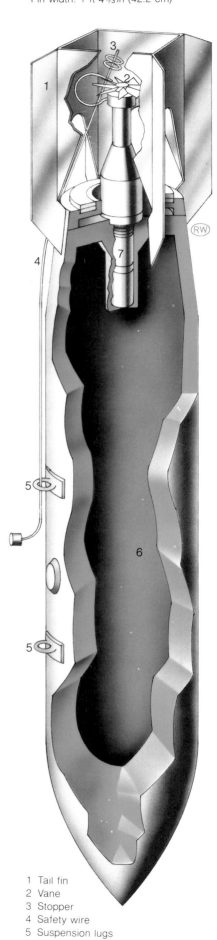

1 Tail fin
2 Vane
3 Stopper
4 Safety wire
5 Suspension lugs
6 Explosive
7 Exploder container

the aircraft; frostbite was always a risk if a bare hand was exposed; and wearing an oxygen mask, sometimes for six hours at a stretch, was not pleasant. Long flights under these conditions were very fatiguing. A failure in the oxygen supply to any crewman could quickly cause unconsciousness and sometimes death.

A major criticism of the Fortress was its small bomb-load in comparison with some other four-engined bombers. With the use of underwing racks the B-17 could, and did on occasions, lift operational loads of 12,000 lb (5,440 kg) whereas its normal internal load averaged between 4,000 and 5,000 lb (1,814 and 2,270 kg). The maximum load that a B-17F could carry was 9,600 lb (4,355 kg), made up of six 1,600 lb (726 kg) armour-piercing bombs, a load that was rarely carried and could only be used on short-range targets. This compared favourably with the maximum for a B-24 Liberator of 12,800 lb (5,806 kg) under similar conditions. A marked contrast was in the actual average operational tonnages carried by RAF Lancasters and 8th Air Force Fortresses to the same target area – 10 and 4 tons respectively. The B-17's limiting factors were the size and construction of the bomb-bay. The aircraft had been designed as a medium bomber. Not only did bomb-bay size limit the overall tonnage carried, but it also restricted individual bombs to a 2,000 lb (907 kg) type maximum. On the face of it, the USAAF's method of taking 4,000 lb (1,814 kg) of bombs to a target appeared highly extravagant in manpower, materials and fuel.

The USAAF argued that the really important factor was the actual destruction wrought at the target, and that their form of precision bombing had a much greater chance of success than the RAF's area bombing at night. Indeed, during the summer and autumn of 1943 8th Air Force Fortresses were able to cause considerable destruction at a number of German war plants. The success of their operations was evident by the efforts the enemy made to thwart the raids, diverting fighters from the Eastern Front and causing losses which threatened to make these daylight raids prohibitive. Priorities then shifted to the introduction of long-range fighters for escort but these did not become available in substantial numbers until the early months of 1944. Meanwhile a notable improvement in the Fortress's forward firepower was made. Efforts to develop a power turret had been concentrated on the Bendix remotely controlled unit which appeared in pre-production form on the YB-40s. Production units became available in late July 1943 and were initially fitted to the last 86 Douglas-built B-17Fs. These later became B-17Gs, the designation used for all chin-turreted Fortresses from Boeing and Lockheed-Vega production. They became the major Fortress production model with the three plants jointly turning out a total of 8,680 during nearly two years of production.

B-17Gs were really an ongoing production development of the B-17Fs, the new designation simply acknowledging the fitting of the chin turret; no other major changes followed worthy of recognition. Significant changes in functioning and performance would follow during the course of production but from a pilot's point of view the early

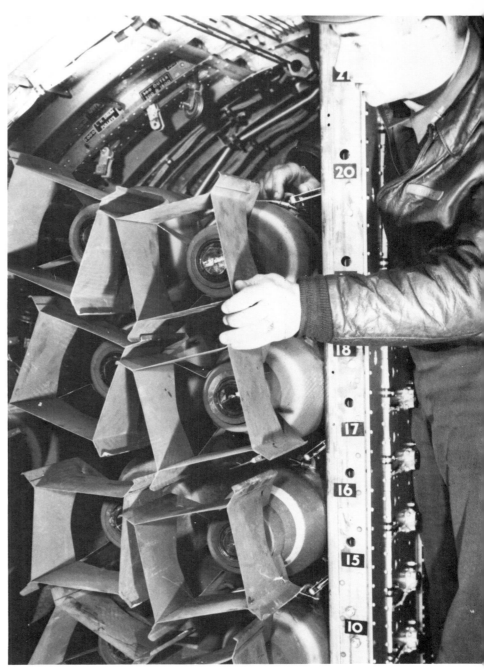

A bombardier checks 100 lb (45 kg) bombs in a B-17 bomb bay. (USAF via G. S. WILLIAMS)

G models were simply late B-17Fs with chin turrets. The new turret was operated by the bombardier through controls on a column which folded back against the starboard side of the nose, so as not to impede his bomb-sighting activities. The computing sight was suspended from the top of the fuselage and projected just inside the Plexiglas nosepiece. There was some criticism of the rate of traverse of these guns and also of their limitations in the field of fire. Fortress manufacturing plants omitted the nose side or cheek guns when the chin turret was introduced. In combat theatres this installation was reinstated for navigators' use and subsequently cheek guns were reintroduced on the production lines.

The first B-17Gs arrived in combat theatres during September 1943 and were soon in action. The B-17F soldiered on for another year, although attrition was such that by the early spring of 1944 groups that had been equipped with the model had only about a half-dozen remaining. The 8th Air Force strength was built up during the winter of 1943–44 to 40 bombing groups, 21 equipped with the B-17 and the remainder with the B-24. Subsequently five B-24 groups were

Angles of defensive fire in B-17G

1 Chin turret: 2 × 0.50-in (12.7 mm) guns
2 Starboard cheek gun: 1 × 0.50-in (12.7 mm)
3 Port cheek gun: 1 × 0.50-in (12.7 mm)
4 Upper turret: 2 × 0.50-in (12.7 mm) guns
5 Lower ball turret: 2 × 0.50-in (12.7 mm) guns
6 Starboard waist gun: 1 × 0.50-in (12.7 mm)
7 Radio compartment gun: 1 × 0.50-in (12.7 mm)
8 Port cheek gun: 1 × 0.50-in (12.7 mm)
9 Tail turret: 2 × 0.50-in (12.7 mm) guns

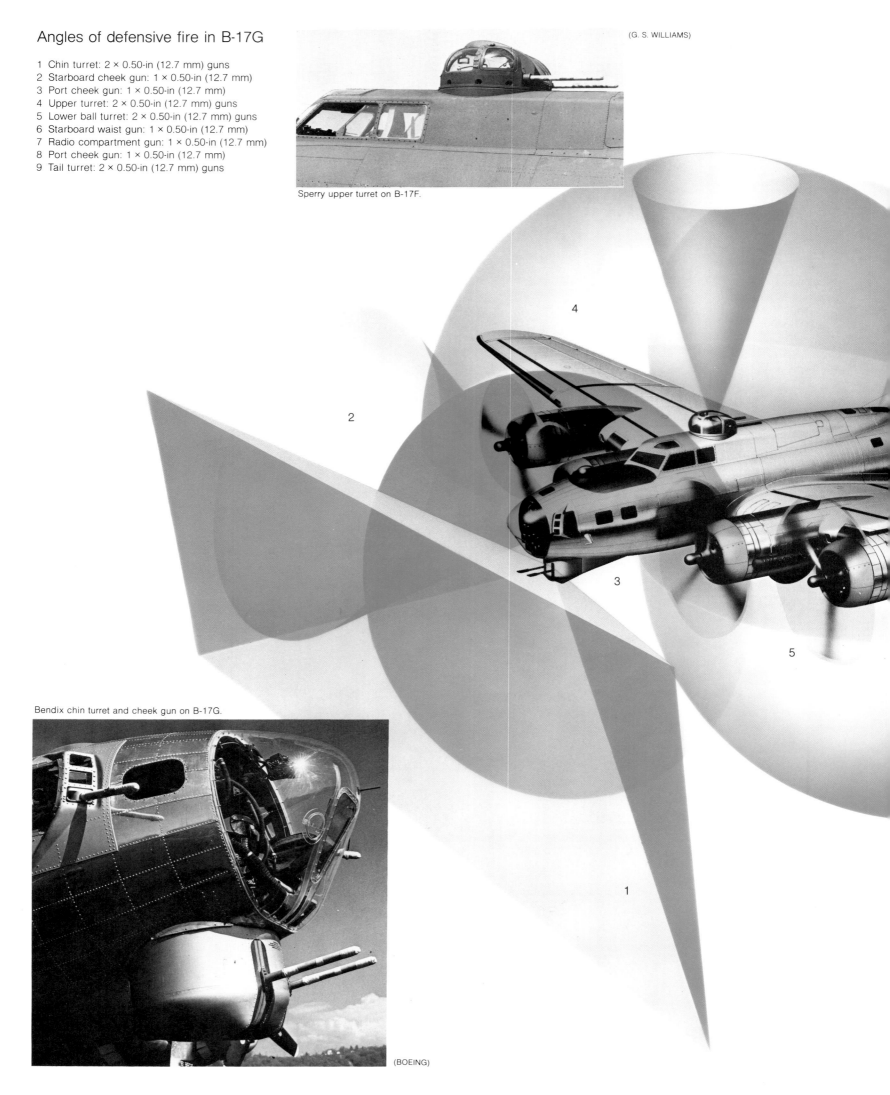

(G. S. WILLIAMS)

Sperry upper turret on B-17F.

Bendix chin turret and cheek gun on B-17G.

(BOEING)

7

9

8

Ball type "Cheyenne" tail gun turret developed by the United Airlines bomber modification centre. It was fitted on late production B-17Gs.

(BOEING)

Sperry ball turret.

(G. S. WILLIAMS)

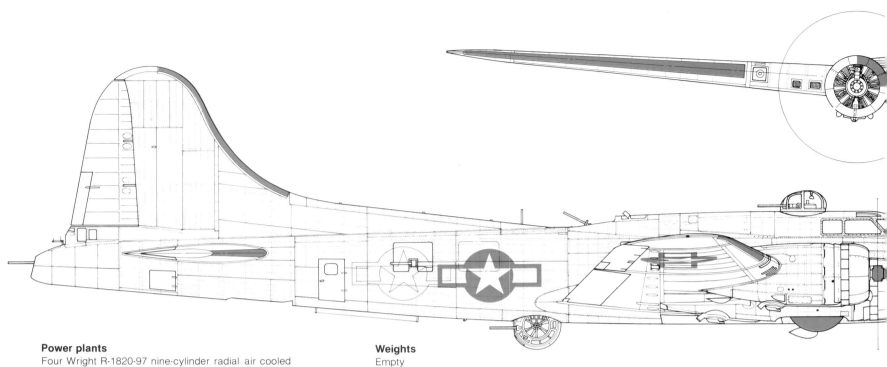

Power plants

Four Wright R-1820-97 nine-cylinder radial air cooled
turbo-supercharged engines
  : 1,200 hp at 25,000 ft (7,620 m)
  : 1,380 hp in "war emergency" power at 25,000 ft (7,620 m).
Hamilton Standard three-bladed hydromatic variable pitch
constant speed fully-feathering propellers
Diameter
  : 11 ft 7 in (3.53 m)
Total fuel in the wings
  : 2,810 US gal (2,340 Imp. gal. 10,637 ltr)
Provision for two overload tanks in bomb-bay
  : 820 US gal (682 Imp. gal. 3,104 ltr)

Performance

Max speed
  : 287 mph (462 km/h) at 25,000 ft (7,620 m)
Max speed in "war emergency" power
  : 302 mph (486 km/h) at 25,000 ft (7,620 m)
Cruising speed
  : 182 mph (293 km/h) at 10,000 ft (3,050 m)
Most economical climbing speed
  : 140 mph (225 km/h)
Max permissible diving speed
  : 305 mph (491 km/h)
Take-off speed
  : 110-115 mph (177-185 km/h)
Landing speed
  : 90 mph (145 km/h)
Time to climb to 20,000 ft (6,100 m)
  : 37.0 min
Service ceiling
  : 35,600 ft (10,850 m)
Take-off distance
  : 3,400 ft (1,040 m)
Landing distance
  : 2,900 ft (880 m)
Range with 6,000 lb (2,722 kg) bomb-load at 10,000 ft (3,050 m)
  : 2,000 mls (3,220 km) at 182 mph (293 km/h)
Max range (ferry) at 10,000 ft (3,050 m)
  : 3,400 mls (5,470 km) at 180 mph (290 km/h)

Weights

Empty
  : 36,135 lb (16,391 kg)
Normal gross
  : 55,000 lb (24,950 kg)
Max take-off
  : 65,500 lb (29,710 kg)

Armament

Bendix chin turret
  : Two 0.50-in (12.7 mm) Browning M2 machine guns with
    365 rounds each
Sperry upper turret
  : Two similar guns with 375 rounds each
Sperry ball turret
  : Two similar guns with 500 rounds each
Tail turret
  : Two similar guns with 500 rounds each
Cheek gun positions
  : Two similar guns with 610 rounds in all
Radio compartment (removed from later aircraft)
  : One similar gun
Waist gun positions
  : Two similar guns with 600 rounds each

Bombs

Normal max six 1,600 lb (726 kg) and two 4,000 lb (1,814 kg)
  : 9,600 lb (4,355 kg) in all
  : 8,000 lb (3,629 kg) in all

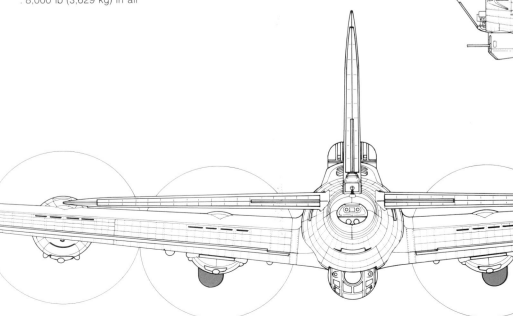

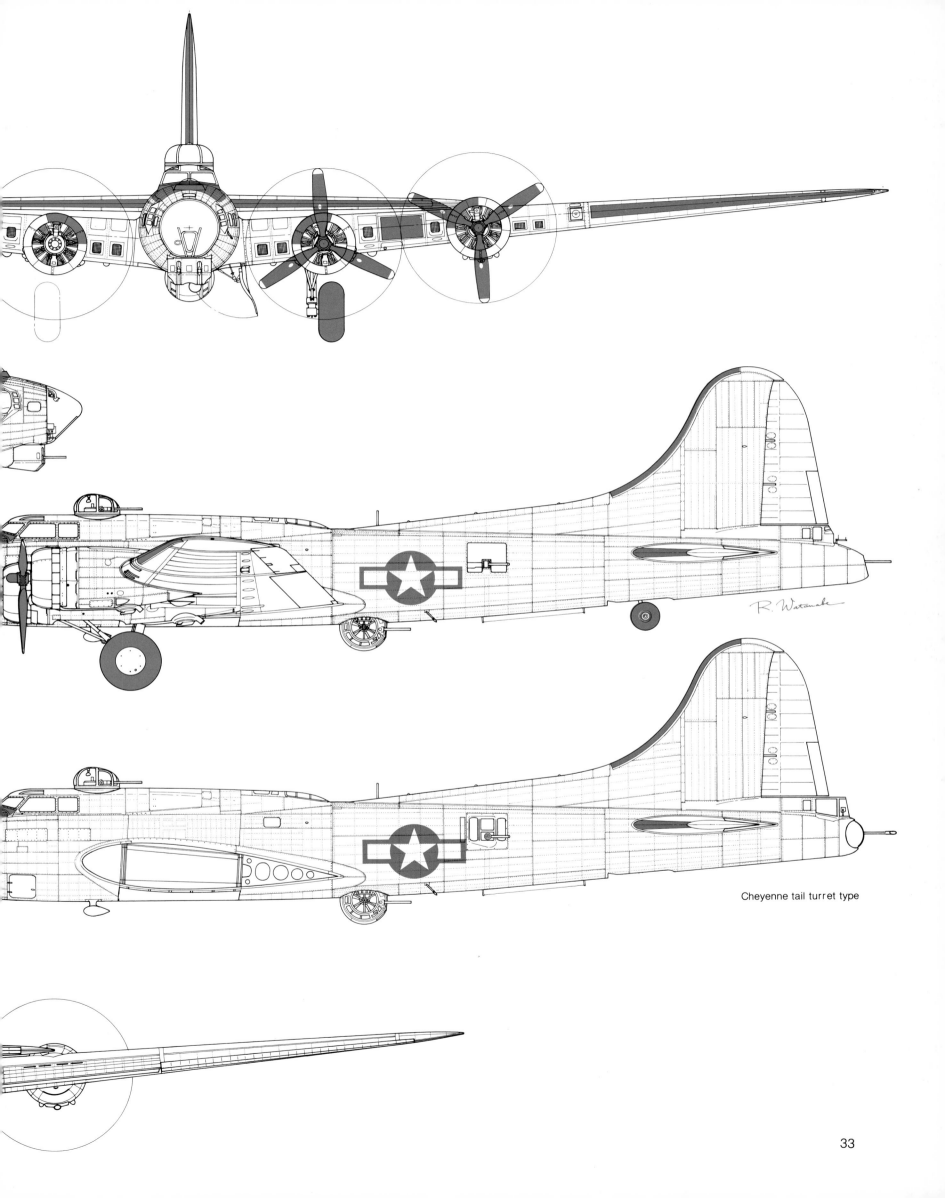

Cheyenne tail turret type

R. Watanabe

33

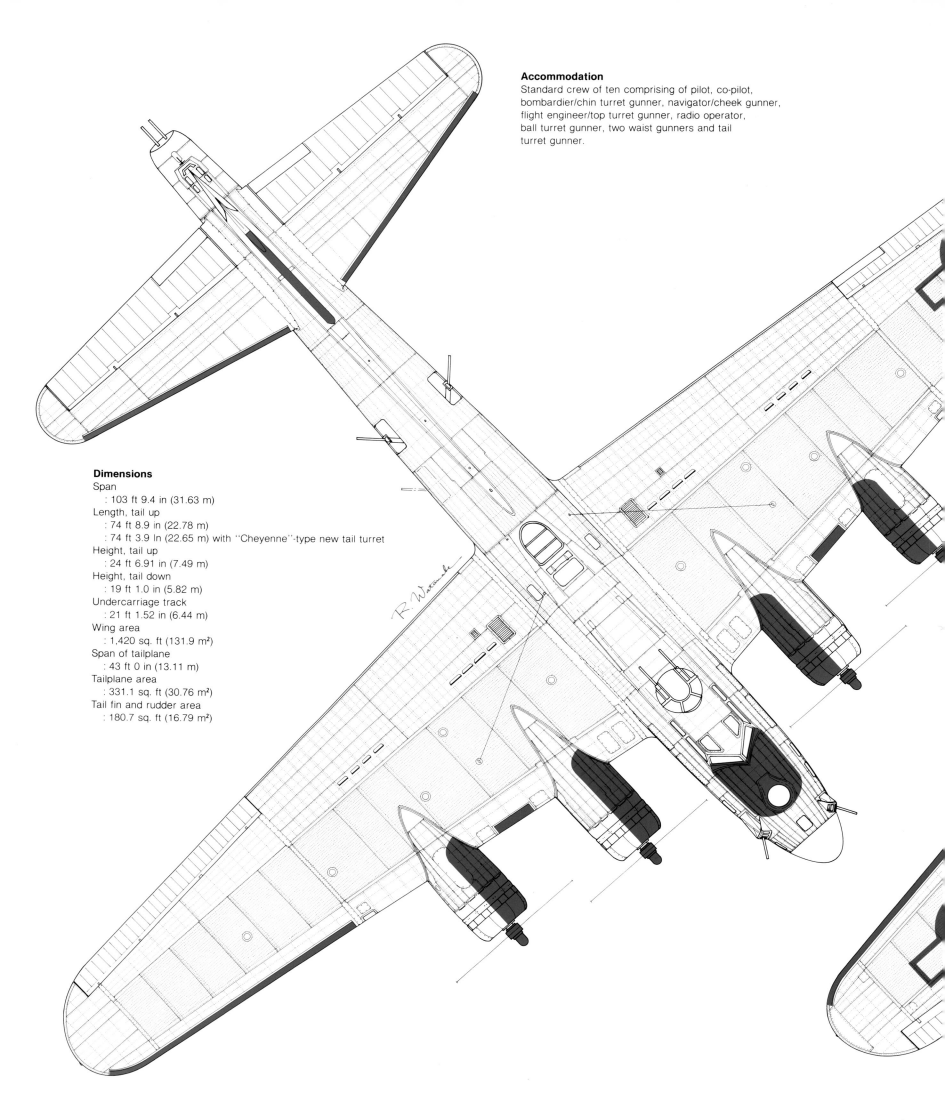

Accommodation
Standard crew of ten comprising of pilot, co-pilot,
bombardier/chin turret gunner, navigator/cheek gunner,
flight engineer/top turret gunner, radio operator,
ball turret gunner, two waist gunners and tail
turret gunner.

Dimensions
Span
 : 103 ft 9.4 in (31.63 m)
Length, tail up
 : 74 ft 8.9 in (22.78 m)
 : 74 ft 3.9 ln (22.65 m) with "Cheyenne"-type new tail turret
Height, tail up
 : 24 ft 6.91 in (7.49 m)
Height, tail down
 : 19 ft 1.0 in (5.82 m)
Undercarriage track
 : 21 ft 1.52 in (6.44 m)
Wing area
 : 1,420 sq. ft (131.9 m²)
Span of tailplane
 : 43 ft 0 in (13.11 m)
Tailplane area
 : 331.1 sq. ft (30.76 m²)
Tail fin and rudder area
 : 180.7 sq. ft (16.79 m²)

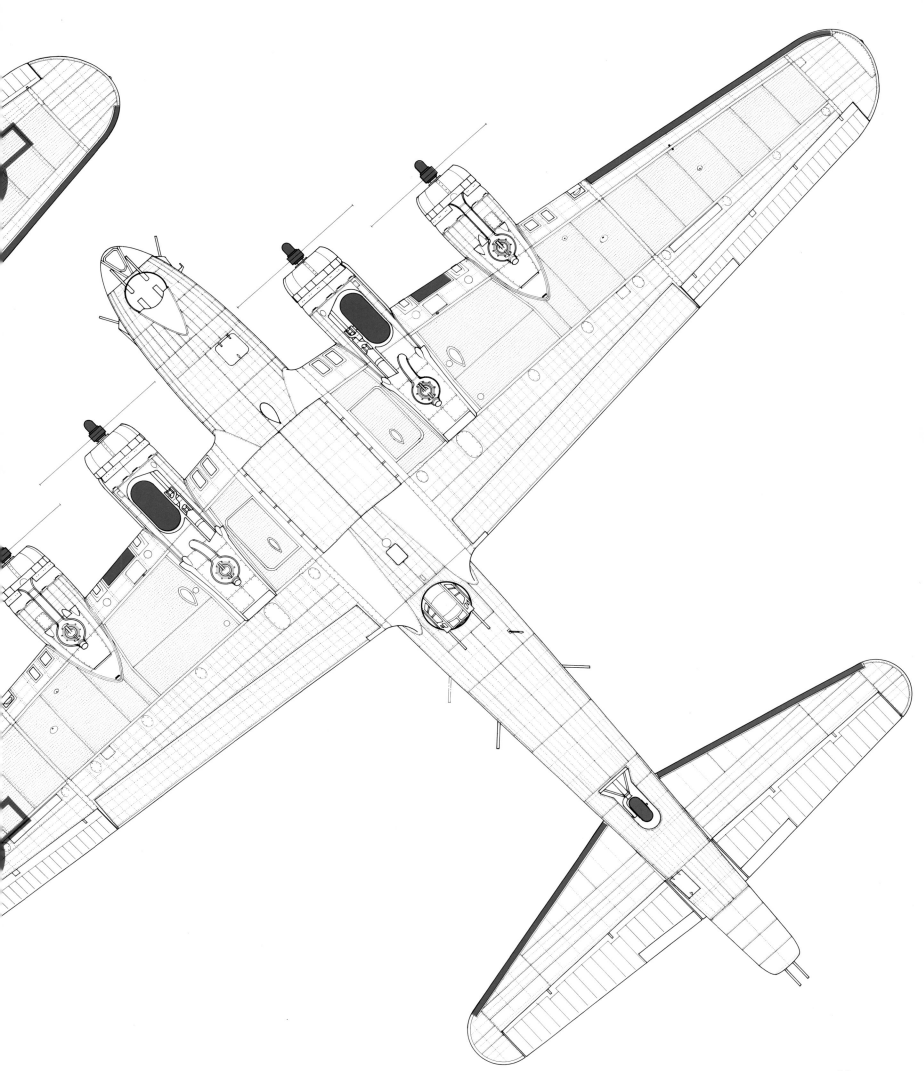

converted to the B-17 to make a total of 26 Fortress groups, which remained in England until the end of the war. Additional units, equivalent to the strength of another combat group, engaged in special activities with the B-17.

In the Mediterranean area a new air force, the 15th, was formed in late 1943 to conduct strategic bombing operations from bases in southern Italy. To the nucleus formed by the four B-17 and two B-24 groups already in that war zone, 15 heavy bomber groups originally earmarked for the 8th Air Force were added to bring the 15th's total to 21, of which six were equipped with B-17s. The Fortress had been completely withdrawn from operations in Pacific war zones in the autumn of 1943 and numbers had been small for many months previous.

There were a number of armament feature improvements during B-17G production and the first during the winter of 1943–44 was the replacement of the Sperry top-turret by a Bendix model. The new turret had a higher profile that looked rather ungainly in comparison with the neat Sperry, but it allowed the gunner better visibility and response to control. Trouble was experienced with fires in these turrets, eventually traced to fraying electrical wiring and oxygen lines. During operational missions the waist posi-

tions were open so that the gunners in the rear of the aircraft were exposed to the icy blast of the slipstream. Also, being directly opposite, the two gunners frequently bumped each other, sometimes dislodging an oxygen connection with serious consequences. Moving the starboard position forward removed this risk and gave each man more room to manoeuvre. Plexiglas coverings were introduced into production and in early 1944 modification centres began installing a framed type cover at waist positions, also fitted to many older B-17s in combat areas.

The radio room gun hatch was also open during combat missions and modification centres devised a change so that this could remain screened with the machine gun installed. A similar installation was made on production with the G-45-DL from Douglas, the G-55-VE from Lockheed-Vega and in the G-80-BO at Boeing. Another important armament change was planned at the same time as the enclosed radio room gun position. This was the "Cheyenne" tail turret, so called because it had originally been developed by the Cheyenne, Wyoming, B-17 modification centre.

The original B-17 tail position featured a tunnel opening at the extremity of the fuselage through which the guns projected, their traverse being restricted to about 30° in

B-17Gs on the assembly line at Boeing plant. (BOEING)

B-17G-20-DL (the first of this block) of 95th BG, US 8th AF heads for a target at Bremen, Germany. (USAF)

both the horizontal and vertical planes. The sighting of these weapons was by a ring and bead placed just outside the gunner's armoured glass panel and linked to the gun movement by a series of levers and bell cranks. Both in layout and operation there had been need for improvement. The Cheyenne conversion gave the gunner a reflector sight and better visibility by enlarging the window area in his position, and a greater field of fire for the guns which were brought closer to him in a pivoting cupola. These turrets were also fitted to a number of older Fortresses at repair depots in combat zones.

By the summer of 1944 the provision of long-range fighter escort had made attacks by Luftwaffe interceptors on 8th and 15th Air Force bombers more the exception than the rule. Many B-17 gunners completed their tour of 30 or more combat missions without ever firing their guns in action. By this time factory, modification centre and battle zone additions to the B-17 plus the maximum fuel, bomb and ammunition loads and other paraphernalia ranging from extra radio equipment to body armour, meant that many of these bombers grossed as much as 30 tons at take-off. The penalty of this overload was reduced range, marred control and engine strain. Concern at the situation led to a reduction in defensive armament in view of the lessened risk of enemy fighter interference.

The value of the radio room gun had always been in question. Here the gunner's outlook was poor and he was rarely able to see the approach of an enemy aircraft; his contribution to defence was usually a hasty burst of fire as an enemy fighter flashed by. The use of this gun position was discontinued by combat groups in Europe and it was eventually omitted on the B-17 production lines. Reducing the quantity of ammunition carried was another measure to cut down the gross weight – during the great air battles of the previous

autumn and winter as many as 10,000 rounds were being carried on some Fortresses. The lessened fighter activity also reduced the number of waist gunners per aircraft from two to one. In the last few months of the war, when enemy fighter attacks were predominantly from the rear quarter (sometimes by fast closing jet aircraft) limited evasive action by the bombers was often considered to be more effective than defensive armament.

In the 8th Air Force an experimental programme of further armament reductions was introduced to lessen weight and drag in order to enhance performance and manoeuvrability. In some units no waist guns or gunners were carried, in others the ball or chin turrets were removed. One group removed chin and ball turrets and cheek and waist guns, increasing top speed by over 20 mph (32.2 km/hr) and giving benefits in rate of climb and endurance. The top speed of a standard B-17G was officially given as 278 mph (447 km/hr). Top speeds, however, had little bearing on the combat performance of the B-17s as the mode of operation entailed flying formations where a specified cruising speed had to be maintained and this varied from 165 to 180 mph (266 to 290 km/hr) IAS (Indicated Air Speed).

With the demise of the German fighter defences came an increase in anti-aircraft artillery fire against the USAAF bomber formations. Flak weapons with controlling radars had been developed to a point where 88 mm and 105 mm shells could be exploded at a selected proximity with great accuracy even 6 miles (9.7 km) up. Some targets were defended by as many as 300 flak guns and the barrages they put up were formidable. Countermeasures were difficult although some success was obtained in jamming the gun-laying radars and by frequent bomber formation deviations in course and altitude. A large amount of armour plate had been added for crew and component protection during the produc-

37

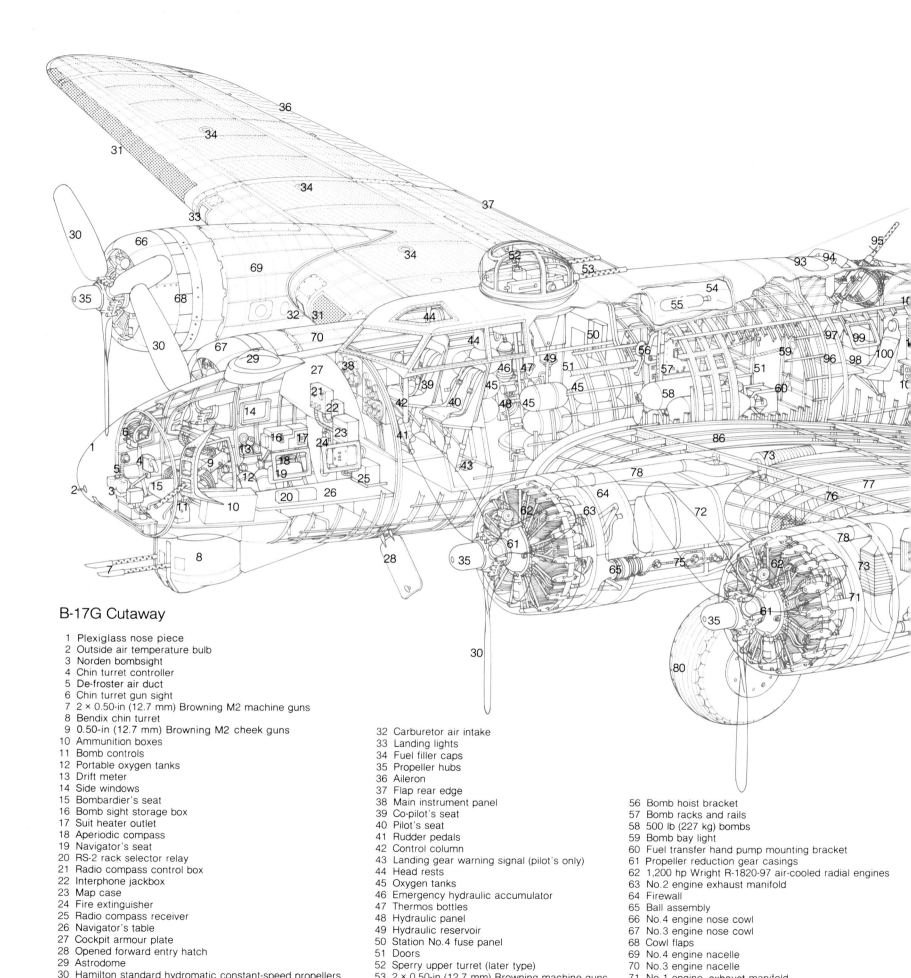

B-17G Cutaway

1 Plexiglass nose piece
2 Outside air temperature bulb
3 Norden bombsight
4 Chin turret controller
5 De-froster air duct
6 Chin turret gun sight
7 2 × 0.50-in (12.7 mm) Browning M2 machine guns
8 Bendix chin turret
9 0.50-in (12.7 mm) Browning M2 cheek guns
10 Ammunition boxes
11 Bomb controls
12 Portable oxygen tanks
13 Drift meter
14 Side windows
15 Bombardier's seat
16 Bomb sight storage box
17 Suit heater outlet
18 Aperiodic compass
19 Navigator's seat
20 RS-2 rack selector relay
21 Radio compass control box
22 Interphone jackbox
23 Map case
24 Fire extinguisher
25 Radio compass receiver
26 Navigator's table
27 Cockpit armour plate
28 Opened forward entry hatch
29 Astrodome
30 Hamilton standard hydromatic constant-speed propellers
   (paddle bladed type)
31 Wing leading-edge de-icer boots

32 Carburetor air intake
33 Landing lights
34 Fuel filler caps
35 Propeller hubs
36 Aileron
37 Flap rear edge
38 Main instrument panel
39 Co-pilot's seat
40 Pilot's seat
41 Rudder pedals
42 Control column
43 Landing gear warning signal (pilot's only)
44 Head rests
45 Oxygen tanks
46 Emergency hydraulic accumulator
47 Thermos bottles
48 Hydraulic panel
49 Hydraulic reservoir
50 Station No.4 fuse panel
51 Doors
52 Sperry upper turret (later type)
53 2 × 0.50-in (12.7 mm) Browning machine guns
54 Left-hand life raft
55 Life raft CO₂ inflation bottle

56 Bomb hoist bracket
57 Bomb racks and rails
58 500 lb (227 kg) bombs
59 Bomb bay light
60 Fuel transfer hand pump mounting bracket
61 Propeller reduction gear casings
62 1,200 hp Wright R-1820-97 air-cooled radial engines
63 No.2 engine exhaust manifold
64 Firewall
65 Ball assembly
66 No.4 engine nose cowl
67 No.3 engine nose cowl
68 Cowl flaps
69 No.4 engine nacelle
70 No.3 engine nacelle
71 No.1 engine exhaust manifold
72 Oil tank (36.9 US gal/30.7 Imp. gal/140 ltr)
73 Intercooler

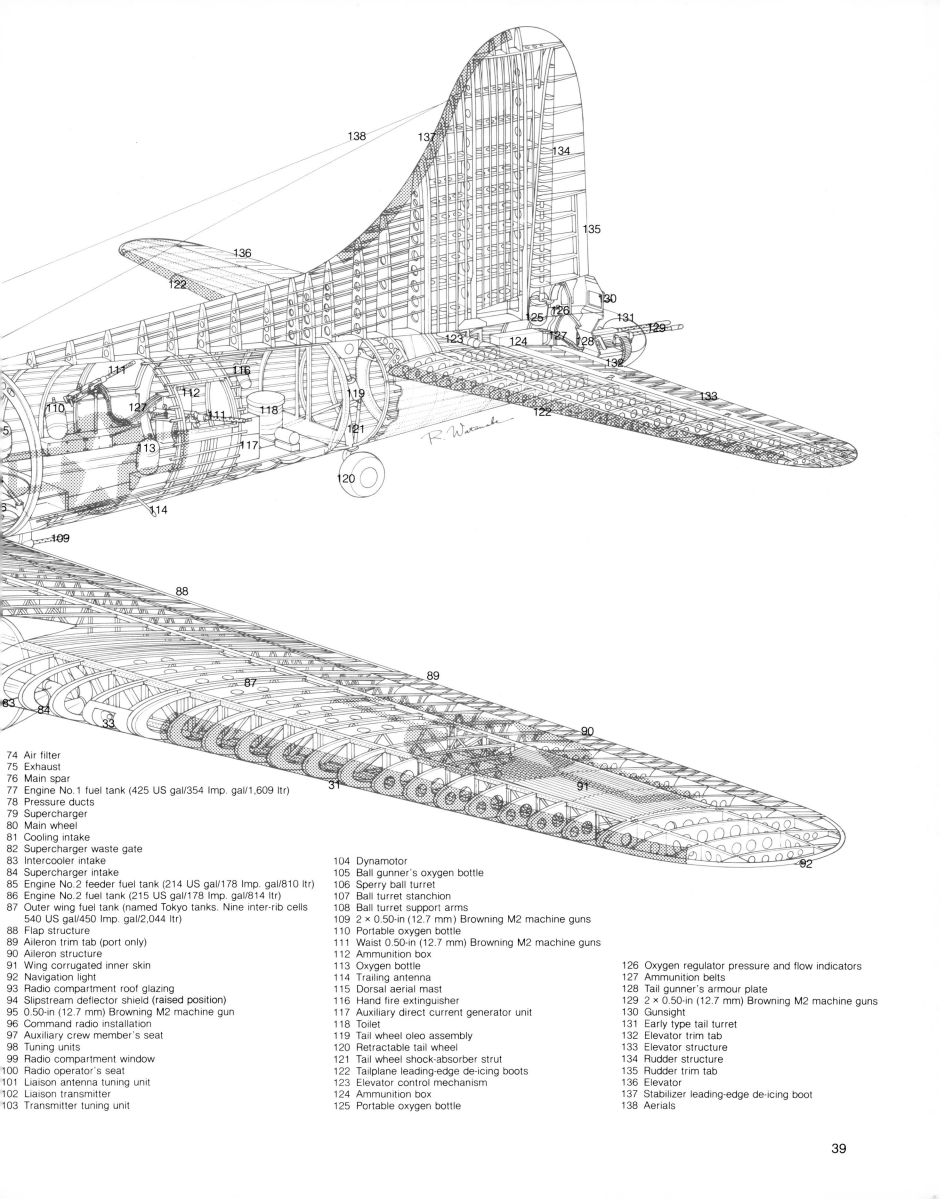

74 Air filter
75 Exhaust
76 Main spar
77 Engine No.1 fuel tank (425 US gal/354 Imp. gal/1,609 ltr)
78 Pressure ducts
79 Supercharger
80 Main wheel
81 Cooling intake
82 Supercharger waste gate
83 Intercooler intake
84 Supercharger intake
85 Engine No.2 feeder fuel tank (214 US gal/178 Imp. gal/810 ltr)
86 Engine No.2 fuel tank (215 US gal/178 Imp. gal/814 ltr)
87 Outer wing fuel tank (named Tokyo tanks. Nine inter-rib cells 540 US gal/450 Imp. gal/2,044 ltr)
88 Flap structure
89 Aileron trim tab (port only)
90 Aileron structure
91 Wing corrugated inner skin
92 Navigation light
93 Radio compartment roof glazing
94 Slipstream deflector shield (raised position)
95 0.50-in (12.7 mm) Browning M2 machine gun
96 Command radio installation
97 Auxiliary crew member's seat
98 Tuning units
99 Radio compartment window
100 Radio operator's seat
101 Liaison antenna tuning unit
102 Liaison transmitter
103 Transmitter tuning unit

104 Dynamotor
105 Ball gunner's oxygen bottle
106 Sperry ball turret
107 Ball turret stanchion
108 Ball turret support arms
109 2 × 0.50-in (12.7 mm) Browning M2 machine guns
110 Portable oxygen bottle
111 Waist 0.50-in (12.7 mm) Browning M2 machine guns
112 Ammunition box
113 Oxygen bottle
114 Trailing antenna
115 Dorsal aerial mast
116 Hand fire extinguisher
117 Auxiliary direct current generator unit
118 Toilet
119 Tail wheel oleo assembly
120 Retractable tail wheel
121 Tail wheel shock-absorber strut
122 Tailplane leading-edge de-icing boots
123 Elevator control mechanism
124 Ammunition box
125 Portable oxygen bottle

126 Oxygen regulator pressure and flow indicators
127 Ammunition belts
128 Tail gunner's armour plate
129 2 × 0.50-in (12.7 mm) Browning M2 machine guns
130 Gunsight
131 Early type tail turret
132 Elevator trim tab
133 Elevator structure
134 Rudder structure
135 Rudder trim tab
136 Elevator
137 Stabilizer leading-edge de-icing boot
138 Aerials

B-17Gs of 99th and 301st BG, US 15th AF release their bombs on the Luftwaffe Me 262 jet fighter base at Lechfeld in southern Germany on 12 September, 1944. (USAF)

tion development of the B-17 but with the change in the form of opposition encountered much of the armour plate was removed at crew positions and so-called flak curtains substituted. These consisted of a series of laminated plates in canvas designed to check low velocity shell splinters.

From a pilot's viewpoint one of the most important improvements to the B-17 was the change from hydraulic to electrically operated turbo-supercharger regulators. The sluggish operation and unreliability of the hydraulic regulator system caused power setting difficulties and added to the problems of flying close formations. The failure of a hydraulic regulator could also precipitate the malfunction of a supercharger. The electrical system gave immediate turbo response, cut down failures and greatly eased control problems in the cockpit. The electric regulators were first installed on production blocks G-10-BO from Boeing, G-15-DL from Douglas, and G-5-VE from Lockheed-Vega. This change was followed a few weeks later by the introduction of a new model turbo-supercharger, the B-22, which through increased turbine speed gave better high altitude performance and improved the B-17's critical altitude. The new turbo came with the G-35-BO, G-35-DL and G-25-VE blocks.

Radar-carrying Models

The main obstacle to high altitude precision bombing was the European weather. Cloud frequently obscured primary targets, forcing bombers to attack targets of opportunity in place of planned targets and sometimes to return with their bombs. To alleviate such frustration, the USAAF turned to the British-developed ground-scanning radar used by RAF night bombers. In the latter half of 1943 twelve installations of this device, known as H2S, were made in B-17FS of the 8th Air Force. The scanning antenna was installed under the nose of the aircraft and covered with a plastic blister reminiscent in shape of the original Model 299 gun blisters. From late September 1943 these radar Fortresses were used as pathfinders for bomber formations, the technique being to fly one in the leading position, the other bombers dropping their bombs on its release signal.

Fortresses equipped with a US-developed version of H2S, designated H2X, commenced operations in November 1943. The first 11 were conversions on Douglas-built B-17G types; the radar scanner, which was partly retractable and

40

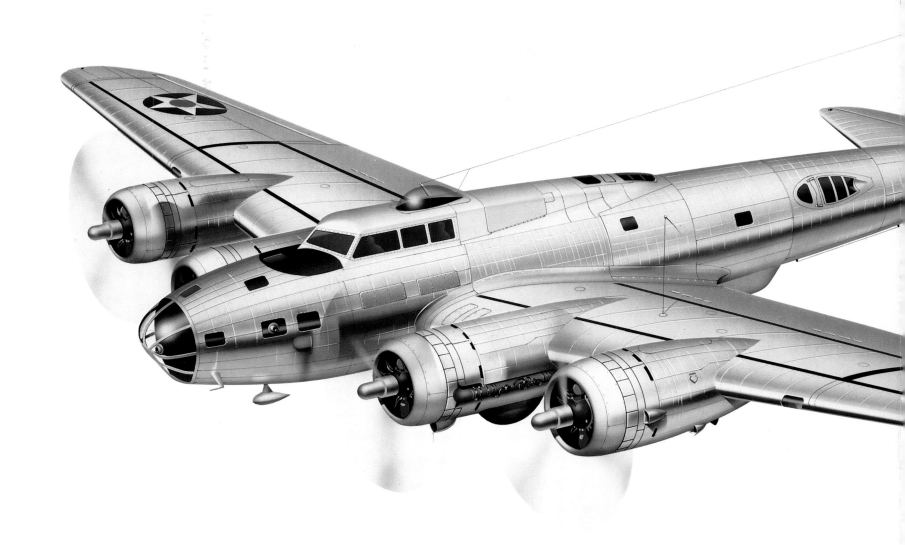

B-17E "Little Skunkface" of 97th BG, US 8th AF in 1942.

Power units
4 × Wright R-1820-65 engines
: 1,200 hp at 25,000 ft (7,620 m)

Performance
Max speed
: 317 mph (510 km/h at 25,000 ft (7,620 m)
Cruising speed
: 195—223 mph (314—359 km/h)
Service ceiling
: 36,600 ft (11,160 m)
Range with 4,000 lb (1,814 kg) bombs
at 15,000 ft (4,570 m)
: 2,000 mls (3,220 km) at 224 mph (360 km/h)

Armament
8 × 0.50-in (12.7 mm) and
1 × 0.30-in (7.62 mm) machine guns
14 × 300 lb (136 kg) bombs (max)

Weights
Empty
: 32,250 lb (14,630 kg)
Max take-off
: 53,000 lb (24,040 kg)

Dimensions
Span
: 103 ft 9.4 in (31.63 m)
Length, tail up
: 73 ft 9.7 in (22.50 m)
Height, tail down
: 19 ft 1 in (5.82 m)
Wing area
: 1,420 sq. ft (131.9 m²)

Crew
10

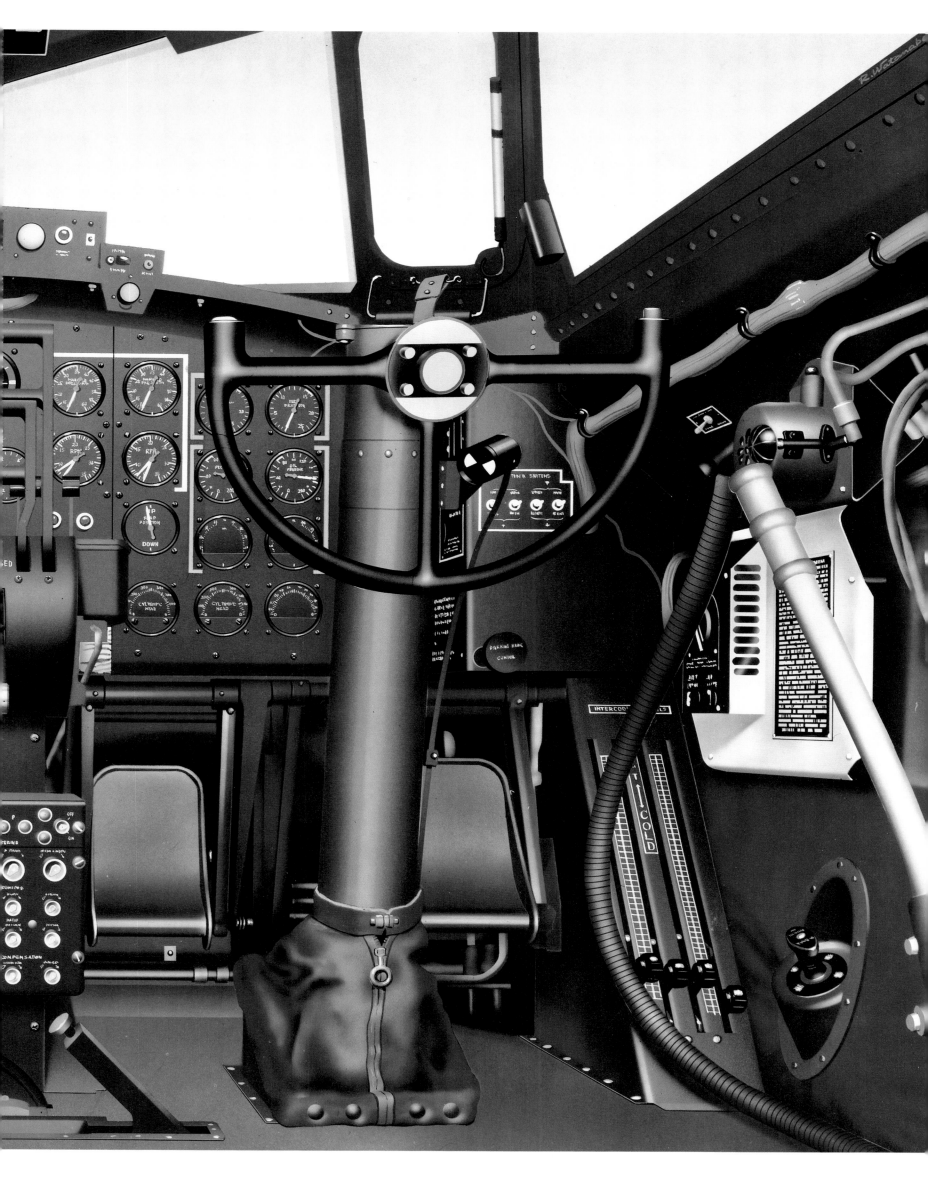

B-17G Cockpit

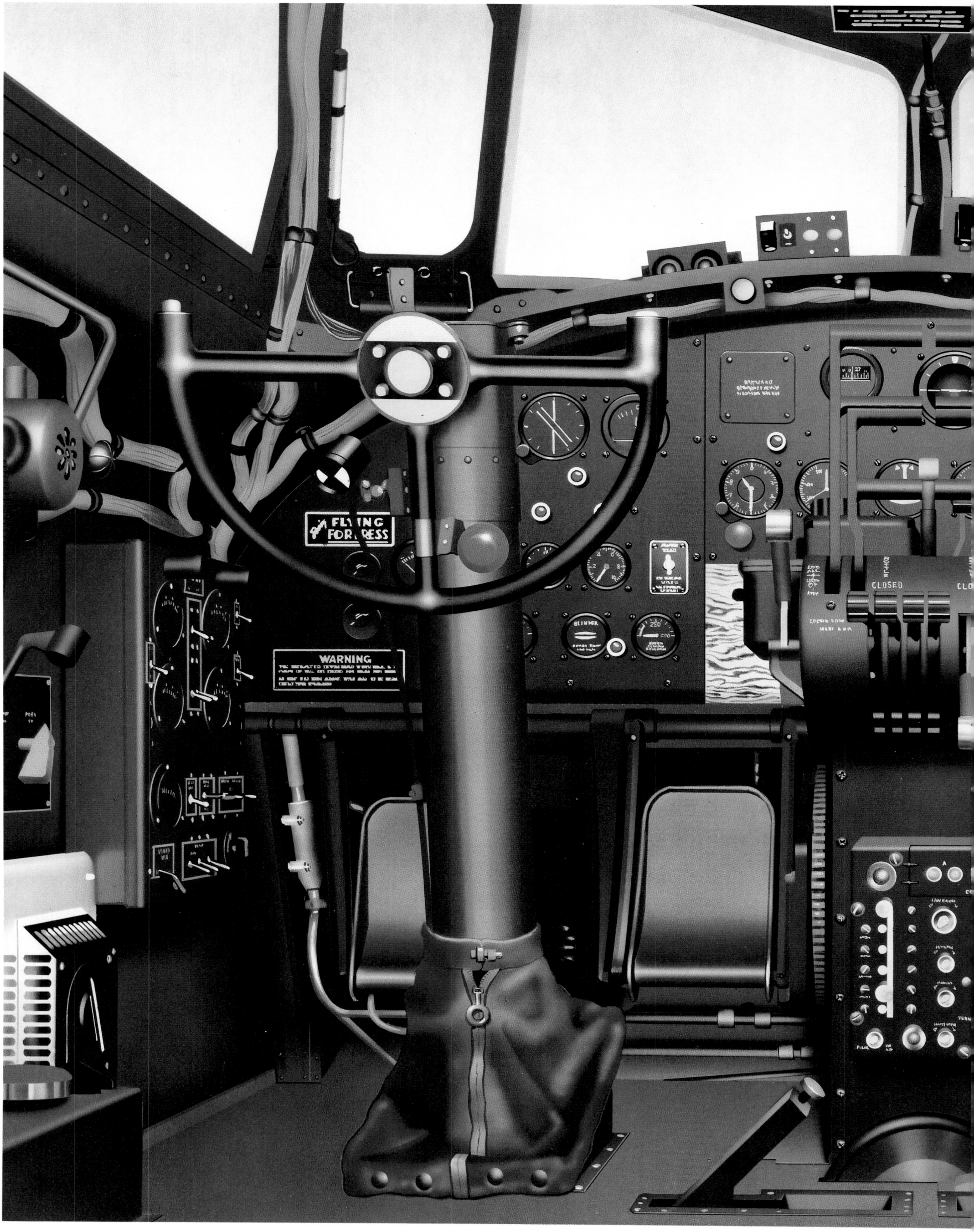

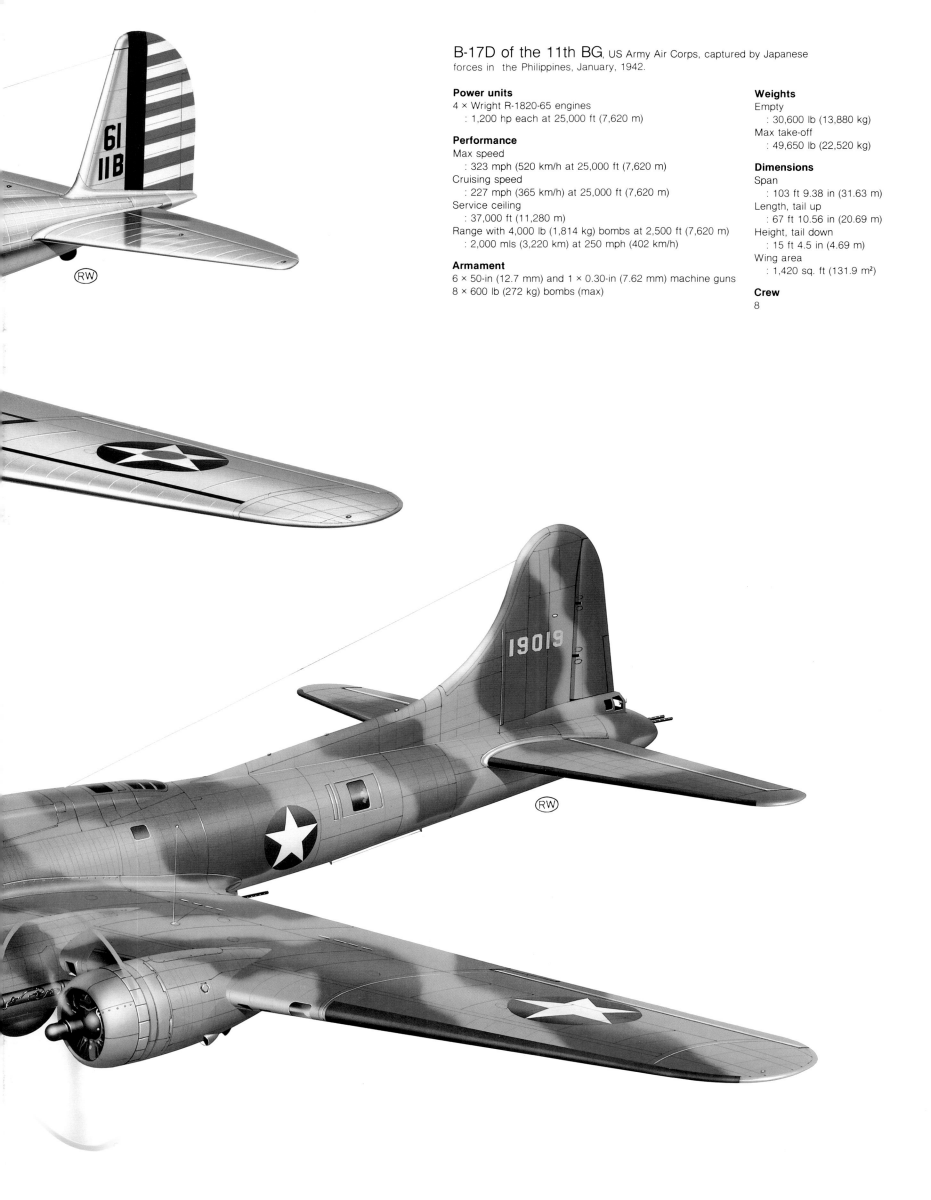

B-17D of the 11th BG, US Army Air Corps, captured by Japanese forces in the Philippines, January, 1942.

Power units
4 × Wright R-1820-65 engines
: 1,200 hp each at 25,000 ft (7,620 m)

Performance
Max speed
: 323 mph (520 km/h at 25,000 ft (7,620 m)
Cruising speed
: 227 mph (365 km/h) at 25,000 ft (7,620 m)
Service ceiling
: 37,000 ft (11,280 m)
Range with 4,000 lb (1,814 kg) bombs at 2,500 ft (7,620 m)
: 2,000 mls (3,220 km) at 250 mph (402 km/h)

Armament
6 × 50-in (12.7 mm) and 1 × 0.30-in (7.62 mm) machine guns
8 × 600 lb (272 kg) bombs (max)

Weights
Empty
: 30,600 lb (13,880 kg)
Max take-off
: 49,650 lb (22,520 kg)

Dimensions
Span
: 103 ft 9.38 in (31.63 m)
Length, tail up
: 67 ft 10.56 in (20.69 m)
Height, tail down
: 15 ft 4.5 in (4.69 m)
Wing area
: 1,420 sq. ft (131.9 m²)

Crew
8

B-17F-25-DL "Winsome Winn" of 381st BG, 534th BS, US 8th AF in July, 1943.

Power units
4 × Wright R-1820-97 engines
: 1,200 hp each at 25,000 ft (7,620 m)
  1,380 hp in "war emergency" power at 25,000 ft (7,620 m)

Performance
Max speed
: 314 mph (505 km/h) in "war emergency" power
Cruising speed
: 200 mph (322 km/h) at 10,000 ft (3,050 m)
Service ceiling
: 37,500 ft (11,430 m)
Range with 6,000 lb (2,722 kg) bombs at 10,000 ft (3,050 m)
: 1,300 mls (2,090 km) at 200 mph (322 km/h) (Tokyo Tanks)

Armament
11 × 0.50-in (12.7 mm) machine guns
8 × 1,000 lb (454 kg) bombs (max)

Weights
Empty
: 34,000 lb (15,420 kg)
Max take-off
: 65,500 lb (29,710 kg)

Dimensions
Span
: 103 ft 9.4 in (31.63 m)
Length, tail up
: 74 ft 8.9 in (22.78 m)
Height, tail down
: 19 ft 1 in (5.82 m)
Wing area
: 1,420 sq. ft (131.9 m²)

Crew
10

7G Bombardier's compartment

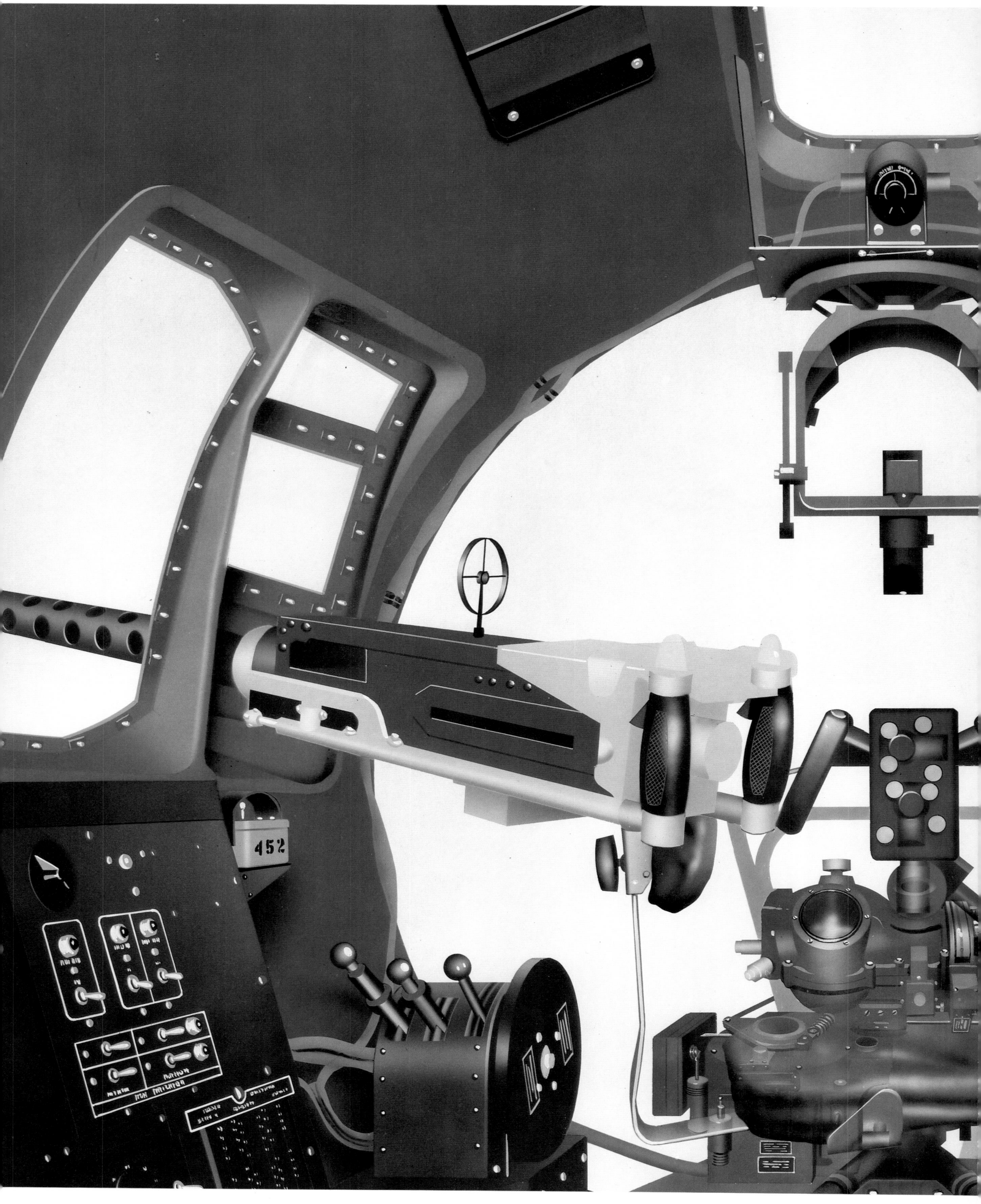

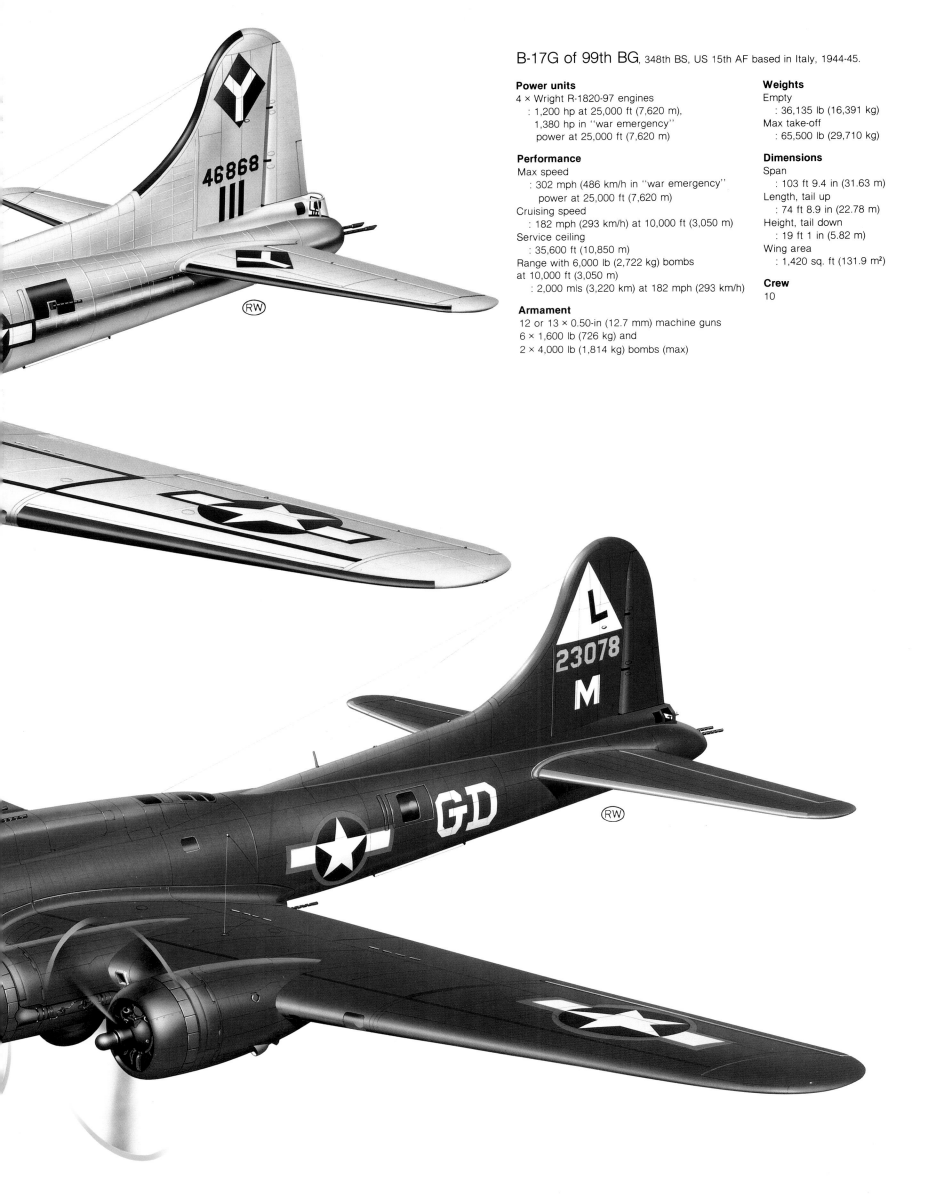

B-17G of 99th BG, 348th BS, US 15th AF based in Italy, 1944-45.

Power units
4 × Wright R-1820-97 engines
: 1,200 hp at 25,000 ft (7,620 m),
1,380 hp in "war emergency"
power at 25,000 ft (7,620 m)

Performance
Max speed
: 302 mph (486 km/h in "war emergency"
power at 25,000 ft (7,620 m)
Cruising speed
: 182 mph (293 km/h) at 10,000 ft (3,050 m)
Service ceiling
: 35,600 ft (10,850 m)
Range with 6,000 lb (2,722 kg) bombs
at 10,000 ft (3,050 m)
: 2,000 mls (3,220 km) at 182 mph (293 km/h)

Armament
12 or 13 × 0.50-in (12.7 mm) machine guns
6 × 1,600 lb (726 kg) and
2 × 4,000 lb (1,814 kg) bombs (max)

Weights
Empty
: 36,135 lb (16,391 kg)
Max take-off
: 65,500 lb (29,710 kg)

Dimensions
Span
: 103 ft 9.4 in (31.63 m)
Length, tail up
: 74 ft 8.9 in (22.78 m)
Height, tail down
: 19 ft 1 in (5.82 m)
Wing area
: 1,420 sq. ft (131.9 m²)

Crew
10

covered by a plastic dome, was positioned directly behind the chin turret. This arrangement caused very cramped conditions in the nose compartment, particularly for the bombardier. Meanwhile, experiments in the US showed that the H2X antenna could be moved further back under the aircraft without spoiling signal transmission or reception. All other H2X-equipped B-17Gs, the first of which arrived in England in late January 1944, had the scanner antenna in a retractable radome placed in the well normally occupied by the ball turret. The radar operator and the display screen (or scope) were situated in the radio room of these aircraft. Initially assigned to the special Pathfinder Group, as more H2X B-17s became available, special units were set up in combat bomber groups. Eventually every B-17 station had a small number of these radar-equipped aircraft for leading missions. The 15th Air Force in Italy also established H2X B-17 units during 1944.

Another form of radar bombing, known as Oboe, was tried out in B-17s in late 1943 but abandoned after only a few trial missions. The system depended upon a signal sent from ground stations in England but this could not be satisfactorily received at high altitude over a long distance. Externally Oboe B-17Fs could only be distinguished from other F models by additional fuselage aerial masts. A very accurate form of short-range radar bombing known as G-H, which reversed the principle of the Oboe system in that the bomber transmitted signals to ground stations in order to obtain its position, was introduced in 8th Air Force B-17s during the spring of 1944. A further development, Micro-H using micro wavelengths, was even more precise and was used with considerable effect during the last six months of the war. Both G-H and Micro-H equipped B-17 lead aircraft and were outwardly only distinguished by additional aerial masts.

A new form of ground-search airborne radar developed in the US was produced in late 1944 and employed during 1945 in special test bombing operations by both 8th and 15th Air Forces. Known by the name *Eagle* it featured an antenna housed in a 1 ft 4 in by 18 ft (40.6 cm by 5.5 m) aerofoil shaped section suspended under the nose of the B-17 giving the appearance of a forward stabilizer. The antenna swept from side to side through approximately 60°, the beam being formed in the forward path of the aircraft in contrast to the revolving 360° sweep of H2X. Using a much higher frequency than the latter, *Eagle*, officially designated AN/APQ-7, gave a clearer presentation of ground images on the operator's scope. In addition to pathfinder radars, the Fortress was also used for specialized work carrying jamming devices for use against enemy radars. This work was later taken on by B-24s which had more internal space to accommodate the bulky transmitters and equipment.

Other Duties

In their offensive role Fortresses carried a variety of explosive and incendiary bombs. External wing racks were used occasionally but mostly for special weapons. The 8th Air Force experimented with a 2,000 lb (907 kg) glide bomb suspended from each wing rack of a B-17, with the object of attacking heavily flak defended targets without the bomber aircraft having to fly through the defences. Used operationally on only one occasion during the spring of 1944, the weapon was found inaccurate and did not justify the effort involved in delivery. More success was achieved with concrete-piercing rocket bombs during the final stages of the war. This British-made missile was used to penetrate U-boat and E-boat shelters on the Dutch and German coasts. Each rocket bomb grossed over 4,500 lb (2,040 kg) and represented the heaviest combat bombing loads carried by B-17s.

Greater quantities of high explosive were aimed at enemy targets using Fortresses in a unique experiment. Under the codename *Aphrodite*, this highly secret operational project used old B-17Fs and Gs as radio-controlled flying bombs. The Fortresses had all armour, armament and most crew equipment removed, and a load of 20,000 lb (9,070 kg) of high explosive was packed into various parts of the fuselage together with a radio control receiver. A volunteer crew of two would fly the aircraft off its English base and put it on course before bailing out prior to crossing the coast. A following B-17 mother aircraft with special radio transmitters would then take over control of the explosive-laden aircraft and guide it towards its target. The radio control system was in an early stage of development and frequently failed with the result that there was little success in hitting a target. One of these drone B-17s exploded in Sweden and another crash-landed in Germany without exploding, so presenting most of its secrets to the enemy. The experiments were terminated in early 1945 and the designation BQ-7 was introduced retrospectively for these Fortresses.

During hostilities USAAF Fortresses were used for purposes other than offensive. A number were converted for photo-reconnaissance under the designation F-9. The cameras varied in number and type but most of the original batch of 16 B-17Fs converted to F-9s carried a total of six situated in the nose, radio room and tail. One squadron equipped with these aircraft was sent to England in late 1942 and moved to North Africa. However, these aircraft could rarely be operated in a combat area without fighter escort and were eventually withdrawn from service. Some 35 other B-17Fs were converted to F-9A and F-9B standard differing only in camera equipment, most being used in non-combat areas for high altitude photography. The F-9C was the designation given to an unspecified number of B-17Gs similarly converted. Although the F-9 photographic designation was retained for some years, some B-17Gs converted for camera work in the latter part of 1944 were not given the F-9C designation. A squadron equipped with photographic B-17Gs carried out mapping surveys in the Middle East, West Africa and Europe during 1945.

The Fortress's excellent high altitude performance led to its employment for weather reconnaissance in the Mediterranean and western European areas. A special weather reconnaissance squadron operated B-17Gs on regular sorties over the Atlantic to collect meteorological information with their special monitoring equipment. Another B-17 squadron was assigned to drop propaganda leaflets over occupied territories and Germany during darkness. The same unit had, during the autumn of 1943, conducted night bomb-

B-17G Cockpit

B-17G Bombardier's compartment

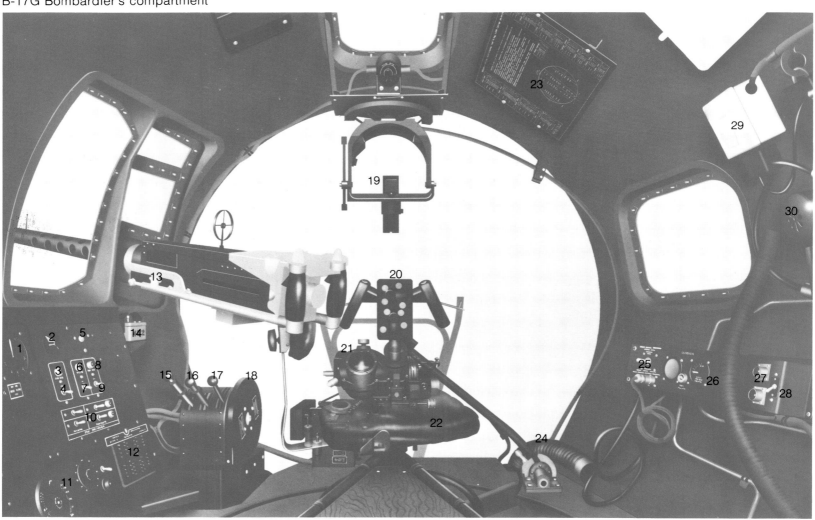

B-17G Cockpit

1 Demand oxygen regulator
2 Cabin air control
3 Vacuum selector valve
4 Suit heater outlet
5 Aileron trim tab control
6 Pilot's control panel
7 Pilot's control panel lights
8 Ammeters
9 Voltmeter
10 Panel lights
11 Fluorescent lights switches
12 Voltmeter
13 Pilot's control column
14 Pilot's rudder pedals
15 Remote compass indicator
16 Pilot's directional indicator
17 Bomb door position light
18 Hydraulic oil pressure gauge
19 Suction gauge
20 Pilot's oxygen flow indicator and pressure gauges
21 Co-pilot's oxygen flow indicator and pressure gauges
22 Co-pilot's oxygen supply warning light
23 Air speed alternate source switch
24 Radio compass position
25 Marker beacon light
26 Altimeter
27 Propeller feathering switches
28 Air speed indicator
29 Turn indicator (Directional gyro)
30 Flight indicator (Gyro horizon)
31 Throttle levers

32 Bank and turn indicator
33 Rate of climb indicator
34 Throttle control lock
35 Propeller pitch controls
36 Propeller pitch control lock
37 Elevator trim tab control wheel
38 Auto flight control panel
39 Elevator and rudder locking lever
40 Rudder tab control wheel
41 Tail wheel locking lever
42 Co-pilot's rudder pedals
43 Co-pilot's control column
44 Tail wheel lock light
45 Landing gear warning light
46 Mixture controls
47 Manifold pressure gauges
48 Tachometers
49 Flap position indicator
50 Cylinder-head temperature gauges
51 Fuel pressure gauges
52 Oil pressure gauges
53 Oil temperature gauges
54 Carburetor air temperature gauges
55 Oil dilution switches
56 Starter switches
57 Parking brake control
58 Intercooler controls
59 Suit heater outlet
60 Demand oxygen regulator
61 Engine primer
62 Hydraulic handpump
63 Interphone jackbox

B-17G Bombardier's compartment

1 Altimeter
2 Ultra-violet light control
3 Bomb formation light
4 Bomb formation switch
5 Bomb indicator light switch
6 Bomb arming light
7 Bomb arming switch
8 Pilot call light
9 Pilot call switch
10 Bomb rack selector switches
11 Intervalometer control panel
12 Bomb indicator
13 Browning M2 0.50-in (12.7 mm) machine gun
14 Bomb release switch
15 External bomb control lever
16 Internal bomb control lever
17 Bomb door control handle
18 Emergency rewind wheel
19 Gunsight setting position
20 Chin turret controller
21 Norden bombsight (phantom view)
22 Bombardier's seat
23 Bomb loading chart
24 De-froster air duct
25 Glide bombing attachment static pressure selector valve
26 Oxygen panel
27 Camera receptacles
28 Camera switch
29 Interphone jackbox
30 Oxygen regulator

A B-17H Dumbo rescue plane (modified from B-17G-105-BO)of 4th Emergency Rescue Squadron, US 20th AF based at Iwo Jima. When photographed, this B-17H was returning from an attempt to save a ditched P-51 pilot with the air-dropped lifeboat. (USAF)

ing experiments with B-17Fs in conjunction with RAF raids on German targets. Fortresses also took part in other night sorties over enemy territory on a very limited scale until the end of hostilities. These were usually to test experimental radar equipment or to obtain photographs of the radar images produced by potential target areas.

The British had successfully developed lifeboats which could be carried under aircraft and dropped by parachute to airmen down in the sea. Late in 1944 a US-developed airborne lifeboat was introduced for carrying under the trusty B-17. The boat was 27 ft (8.2 m) long and

could be carried by any Fortress providing the lower area of the bomb-bay was reworked to accommodate attachment. Early in 1945 a few B-17Gs in England were modified to take airborne lifeboats and used by the Emergency Rescue Squadron operating over the North Sea. During that year some 50 B-17Gs were specially modified in the United States for lifeboat use and several despatched to Pacific war zones. B-17s with airborne lifeboats operated regular patrols from the island of Iwo Jima during the B-29 Superfortress raids on Japan and made a number of drops to the crews of bombers forced down in the sea. Fortresses converted to this air-sea

51

Modified from B-17G-VE (similar to USAAF B-17H). Fitted with an airborne
lifeboat. (US Coast Guard)

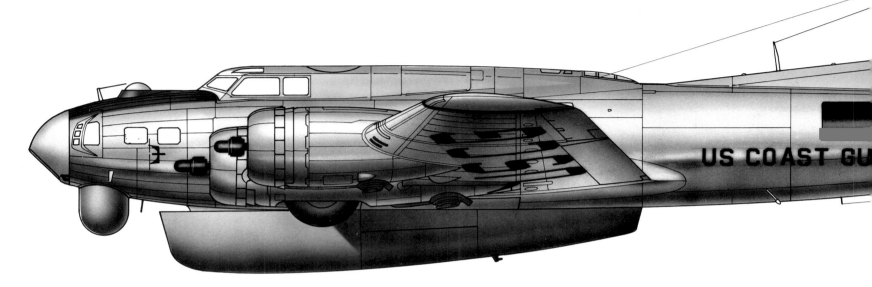

rescue configuration were later given the identity B-17H but,
as far as is known, the majority of wartime conversions operated under their original service designations.

The good handling qualities of the B-17 led to many
war weary bombers retired from combat being converted into
staff transports, air ambulances, target towers or general
"hacks". The few B-17s remaining in the Pacific theatres of
war after the type had been withdrawn from bomber units
during 1943 were all converted for staff transport use. The
most celebrated example was a B-17E specially for use by
General Douglas MacArthur, commander of the US forces in
the South-west Pacific. This aircraft had all camouflage and
remaining armour removed. The radio equipment and
operator were repositioned to the rear of the cockpit, the
bomb doors were sealed and the bay was used as a storage
compartment; the waist windows were also sealed, and the
forward waist area was fitted out with five airline-type seats.
The original door was discarded in favour of a swing-down
type with steps. Additional small windows were fashioned in
the fuselage, a galley with cooking and refrigeration facilities
was fitted out in the former radio room, and a streamlined
fairing was placed over the former tail gun position. This
officially sanctioned conversion, carried out in 1943, was
designated XC-108 and a similarly reworked B-17F, serving the
US Command in Europe and the Middle East, received the
designation YC-108. Neither aircraft had any experimental or
test programme before being despatched to their combat
areas. Stripped of war equipment and having smooth contours, these aircraft were potentially the fastest Fortresses
flying. In practice they were operated at altitudes where
oxygen was not required for crew and passengers and the top
speed attained was only about 300 mph (483 km/hr).

Further investigation of B-17 transport possibilities
resulted in two more conversions under the C for Cargo
designation, XC-108A and XC-108B. The former was a B-17E in
which the interior of the fuselage from the bomb-bay aft was

stripped to allow the maximum in storage space. A large
upward folding cargo door was constructed in the left-hand
side of the fuselage just aft of the wing trailing edge, and the
radio and navigation positions were relocated behind the
pilots. The XC-108B was a B-17F turned into a flying fuel
tanker. Fuel cells were fitted into the bomb-bay area to give a
thousand gallon load. The restrictions imposed by the B-17's
comparatively narrow fuselage made both the cargo and
tanker adaptations uneconomical. The B-24 Liberator, with
its deeper fuselage and more rectangular section, was better
suited to such conversions, in the absence of a sufficient
number of cargo transports. While this investigation of the
B-17's transport potential was taken no further in the US, a
considerable number of B-17s had been and were employed in
transport work. As early as the autumn of 1942 stripped

Fortress IIA

Modified B-17E armed with a 40 mm Vickers "S" gun. This aircraft was at one time
operated by No. 220 Squadron, RAF Coastal Command, over the Central Atlantic.

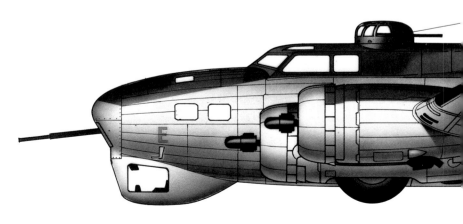

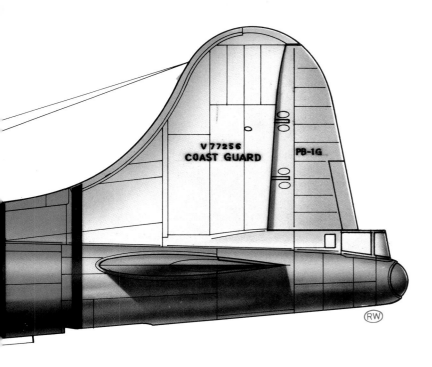

down B-17Es had been used to move supplies from Australia to New Guinea in the absence of sufficient true air transports. Both in Italy and England air depots carried out conversions, removing armament and bombing equipment and installing seating. Although fuselage space and access were restricted, Air Forces Services Command employed lightened Fortresses on a regular basis to carry spares and equipment to combat bases.

In Allied Service

When hostilities terminated in Europe in May 1945, the USAAF had 133 squadrons equipped with Fortresses in that theatre. The Royal Air Force and Royal Canadian Air Force were the only other services with the

Fortress in regular use at this time, the RAF having six squadrons and the RCAF one partially equipped. Despite the unsuccessful use of the Fortress I by RAF Bomber Command in 1941, Britain was scheduled to receive a substantial number of improved B-17F Fortresses for bombing use during 1942 as the Fortress II. America's entry into the war, and the USAAF's desperate need for aircraft to meet its own expansion, brought a revision of plans, and a large number of B-17Es and B-17Fs complete with RAF markings and serial numbers reverted to the USAAF. Later it was decided to supply 45 B-17Es from earlier production to the British and, as these differed from the original planned allocation, they were designated Fortress IIA. These aircraft arrived in Britain during the early summer of 1942 and, as the USAAF was in the process of establishing its own bombing organization equipped with Fortresses in the United Kingdom, the RAF decided to use them for maritime reconnaissance and anti-submarine work.

The RAF eventually received 19 B-17Fs from the earlier contract, and these Fortress IIs reached them in the early autumn of 1942. Both versions were used by RAF Coastal Command. Equipped with submarine-detecting radar, these aircraft carried out oceanic patrol from airfields in the north-west of the UK and in the autumn of 1943, when agreement was reached with the Portuguese Government, two squadrons moved to the Azores. The Command's Fortresses made a number of successful attacks on U-boats during their service before the longer-ranged Liberators replaced them the following year. Remaining Mk II and IIA Fortresses were then used by squadrons in the meteorological role.

The RAF selected the Fortress as the vehicle for a number of test installations, perhaps the most notable being a 40 mm Vickers "S" gun. This was the largest calibre weapon installed in a wartime Fortress – there were a few experimental installations of 20 mm cannon by the USAAF. A gondola had to be constructed below the Fortress's nose to

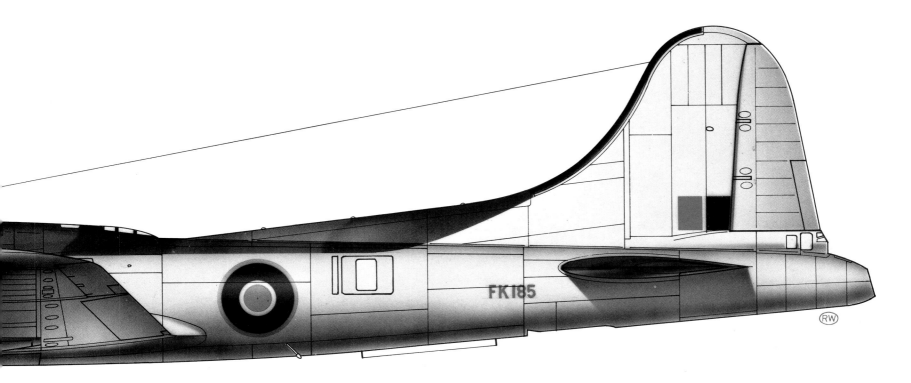

accommodate the gunner who sighted for and fired the 40 mm cannon. Tests revealed unacceptable vibration when the gun was fired and the project was discontinued. Britain did not acquire further production Fortresses until the summer of 1944 when 85 B-17Gs, to be known as Fortress IIIs, were earmarked for delivery. Only 69 actually reached the RAF in Britain, the other 16 apparently being diverted to the USAAF. It is probable that deletions were made because the RAF had received a similar number of Fortresses (B-17Fs) direct from the USAAF in the United Kingdom, most during January 1944. These aircraft were acquired for use in a special radar-jamming squadron and were selected because of the Fortress's superior high altitude performance and the depth of the bomb-bay which could more easily accommodate the bulky jamming transmitters.

In service these lately acquired Fortress IIIs also fitted H2S ground-scanning radar with the antenna housed in a plastic fairing under the nose very similar to that of the early 8th Air Force H2S-equipped pathfinders. Fortress IIIs, similarly modified and with chin turrets removed, were used as replacements in the two radar countermeasures squadrons forming part of RAF No. 100 Group supporting RAF Bomber Command night operations. Various radio and radar countermeasures equipment was carried and these Fortresses usually displayed a number of aerial masts along their fuselages. A few examples of the RAF's B-17G acquisition ended up in Coastal Command meteorological squadrons but for the most part the type was used by RAF Bomber Command.

Canadian squadron use of the Fortress was confined to six aircraft converted for transport duties. In November 1943 three B-17Es and three B-17Fs which had seen service at US training bases were purchased for use in a transatlantic airmail service. Assigned to the RCAF's 168 Heavy Transport Squadron, the aircraft were employed on regular transatlantic flights over the next two years, during which time one disappeared at sea, two were written off in crashes in Europe and another at home base in Canada.

Enemy Fortresses

An unexpected operator of Fortresses during the Second World War was the Luftwaffe. Most combatant nations managed to capture examples of their enemy's warplanes and there were usually flown for evaluation purposes. The Japanese acquired a number of disabled B-17C, D and E models when they overran the Philippines and Java. Subsequently they rebuilt three examples and tested them in Japan. The Germans obtained their first Fortress in December 1942 when a battle-damaged B-17F of the 8th Air Force made a wheels-down forced landing in northern France during a bombing mission. This aircraft was duly evaluated and used to develop effective fighter tactics against the US bomber formations. Two other B-17Fs and a B-17G were acquired in later months. The Luftwaffe, impressed with the high altitude performance and useful range of the B-17, employed the aircraft in clandestine operations behind the Allied lines. KG 200, the organization involved in this work, used Fortresses for agent and equipment deliveries to the Middle East and North Africa. These activities came to the Allies' notice when a Luftwaffe-operated Fortress with mechanical difficulties made a crash-landing in Spain on return from one of these secret operations.

During 1944 the Luftwaffe secured other Fortresses, two when the crews thought they were landing in neutral or Allied-controlled territory. KG 200 extended its operations to liberated France and the UK after the Allied cross-Channel invasion, and it was during a night sortie over Brest to drop supplies to the beleaguered German garrison that the only enemy-operated B-17 to be lost in action was shot down by a British night fighter. At least six different B-17s are known to have been operated by KG 200 which referred to them by the cover name of Do 200. Three are believed to have been destroyed on the ground by Allied air attacks and one was recaptured by US forces at the end of the war.

Postwar Employment

The last of the 12,731 B-17s built was delivered by the Lockheed-Vega plant on 29 July, 1945. Prior to that production at Boeing had been terminated in mid-April of that year and by Douglas in June. Of this total some 5,000 Fortresses had been lost in operational sorties and another 2,000 had been written off as a result of crashes or damage beyond economical repair. Around 2,500 Fortresses returned from Europe at the end of the war were sent to storage depots. Most would never fly again, ultimately to be broken up for salvage. A similar fate awaited the great majority of warplanes but the Fortress's military service was far from over. While the B-29 Superfortress would form the basis of the USAAF's (and later the USAF's) bomber force during the immediate post-war years, the B-17 was relegated to a wide variety of tasks, indicated by the various designations bestowed upon them.

CB-17s were passenger transport conversions which rated as VB-17s if they were to a de luxe standard for high ranking and other VIPs. TB-17 identified a Fortress used for training. FB-17 was a designation for those aircraft fitted out for camera work which, to some extent, clashed with the RB-17 designation for Fortresses with similar equipment in reconnaissance units. In any event, the FB-17 tag apparently faded out while the RB-17 endured. Weather-reconnaissance Fortresses collected the appropriate W prefix and WB-17Gs survived in USAF service well into the 1950s. The model designation B-17H, given to lifeboat carrying B-17Gs in 1945, became identified later as SB-17s (the S standing for search).

Following on the operational trials carried out with radio-controlled Fortresses from England during the war, considerable experimentation in this art at United States' establishments was based on the same reliable airframe. Radio-controlled Fortresses, whatever their mission, were distinguished as QB-17s from 1948 and the controlling aircraft as DB-17s, the D standing for director. MB-17 was a little-used designation applied to Fortresses used to launch guided missiles in the early post-war years. The Fortress in its QB-17 form was to become the first choice for a real target on which to test

A crewless QB-17 Drone starts 2,600 mile flight from Hilo Naval Air Station, Hawaii, to Muroc Army Air Base on 6 August, 1946. This QB-17 was a modified B-17G-110-VE and was used in the Bikini nuclear bomb tests.

the growing assortment of air-to-air and ground-to-air missiles. Radio-controlled B-17s were also used in the Bikini Atoll atom bomb tests of 1946–47, to provide information on blast turbulence and radiation.

Perhaps the most unusual of the conversions were the three five-engined Fortresses. The first of these were the two JB-17Gs, flying test beds for engines, with an additional engine under test mounted in the nose; the flight deck was in consequence repositioned further back. One JB-17G was transferred to Curtiss-Wright to flight test the Wright XT-35

experimental turboprop engine. Later this aircraft was used to test the Wright XJ-65 experimental turbojet mounted under the nose. The other JB-17G, operated by Pratt & Whitney, tested their trial XT-34 turboprop engine. In the early 1950s a third B-17G became a flying test bed. This involved the Allison T-56 turboprop but the installation was made without repositioning the cockpit. In this way the Fortress played its part in bridging the gap to the jet age.

The last Fortress in regular squadron service with the USAF was an SB-17G of the 57th Air Rescue Squadron

Israeli Air Force's B-17Gs flying over the Mediterranean.

based in the Azores until withdrawn in 1956. DB-17 and QB-17 Fortresses continued to be used in missile firing tests until mid-1960.

In 1945 the US Navy had acquired 48 Douglas and Lockheed-Vega-built B-17Gs, 31 of which were fitted out for radar early warning duties under the designation PB-1W. They featured an APS-20 scanner in a large radome under the fuselage centre-section and extra tankage was installed. The remainder of those acquired went to the US Coast Guard, administered by the Navy, under the designation PB-1G. Among their tasks was that of long-range iceberg reconnaissance. The last Coast Guard-operated aircraft, one specially modified for aerial survey work, was not withdrawn until October 1959.

While Britain discarded its Fortresses soon after the war ended and returned the survivors to the US under Lend-Lease agreements, some foreign air forces purchased the type for service. The newly formed state of Israel acquired three B-17Gs, to the apprehension of the Arab world, basing them at Ramat David where they formed the nucleus of a bomber unit for the Israeli Defence Force. The Rio Pact of Mutual Defence of 1947 opened up South American republics as markets for US war surplus armaments, including Fortresses. Dominica used a few B-17Gs ostensibly in the bombing role and as late as 1958 eight B-17Gs were taken out of store in the US to equip a bombing squadron in Bolivia. Chile also acquired B-17Gs but used them in the search-and-rescue role and Brazil bought SB-17Bs for this purpose. In mid-Atlantic the Portuguese for many years stationed three SB-17Gs for long-distance sea search and rescue.

During the Second World War nine B-17Fs and 60 B-17Gs had landed or crashed in Sweden or in Swedish territorial waters. Over 30 of these that were undamaged or only slightly damaged were returned to the USAAF during May and June 1945, but eight others in good condition were bought for commerical use by the Swedish ABA airline to run a transatlantic service. In this venture they were partnered by the Danish line DDL, also operating an ex-USAAF Fortress from the same source. As Denmark had been under German occupation, Fortresses crashed or force-landing in Denmark (nearly 50 in all) were salvaged by the Luftwaffe. DDL sold their Fortress to the Royal Danish Air Force who used it for communications duties with Greenland.

The French *l'Institut Géographique National* based at Creil, near Paris, operated a small number for over 30 years in high altitude photographic mapping; altogether a dozen B-17Gs were registered in France for civil use. The Fortress was preferred for photographic work because it did not have fuselage pressurization, allowing cameras to be used from open hatches and avoiding the risk of distortion posed by window glazing. Fortresses used by a forest fire-fighting company for the aerial application of borate in fire suppression were operated until the early 1980s. By this date remaining airworthy Fortresses were mostly in the hands of air show enthusiasts who flew the veteran aircraft for the delight of spectators.

Boeing have produced many other fine bomber aircraft and become the world's premier manufacturers of commercial transport aircraft bringing the company international respect. Many of their products are renowned for reliability and durability, but it is highly unlikely that any will ever eclipse the fame of the venerable B-17 Flying Fortress.

B-17G-DL of 401st BG, US 8th AF, on final approach to Deenethorpe aerodrome, England.